Good To Go? Decarbonising Travel After the Pandemic

PERSPECTIVES

SERIES EDITOR: DIANE COYLE

Good To Go? Decarbonising Travel After the Pandemic

David Metz

LONDON PUBLISHING PARTNERSHIP

Published by London Publishing Partnership
www.londonpublishingpartnership.co.uk

Published in association with
Enlightenment Economics
www.enlightenmenteconomics.com

ISBN: 978-1-913019-61-7 (pbk)
ISBN: 978-1-913019-62-4 (iPDF)
ISBN: 978-1-913019-63-1 (epub)

A catalogue record for this book is
available from the British Library

This book has been composed in Candara

Copy-edited and typeset by
T&T Productions Ltd, London
www.tandtproductions.com

Printed and bound in Great Britain
by Hobbs the Printers Ltd

www.carbonbalancedprint.com
CBP2250

Contents

Introduction

The coronavirus pandemic has been a defining event of our times. This book is about whether and how we can achieve a positive legacy, rather than letting a good crisis go to waste. The pandemic has been hugely disruptive to travel. The risk of infection in workplaces, in entertainment venues, on public transport and aboard planes has meant that we have travelled much less. This disruption to established travel patterns points to how our travel could change to ensure the transport system reduces its impact on global warming. Future travel must be sustainable.

So how much of our travel is really necessary? And will a combination of less travel and new technologies make travel sustainable?

The modern transport system, as it has developed over the past two centuries, has transformed our lives. The ability to travel at faster than walking pace, particularly by car, has hugely enlarged our horizons. We now need to ask whether the experience gained during the pandemic might make us willing to move at a slower pace and travel less in order to mitigate the environmental impact of our mobility, accepting some loss of access benefits as a consequence.

One objective that governments had at the start of the pandemic was to decarbonise faster by promoting walking and cycling, but this failed to prevent a rebound in car use. There are many ways of cutting transport sector carbon emissions, and a key question is whether reducing car use by changing people's behaviour *en masse* is essential, or whether we can rely on technological developments – electric vehicles in particular – to achieve our goal.

The broad conclusion of this book is that while there is some scope for travelling less to work and for shopping, we are unlikely to voluntarily accept less access to the people and places we have become used to. We will therefore want to rely largely on technology to cut transport carbon emissions, but will that happen fast enough to meet our climate change objectives? And if not, will we need additional incentives to induce significant behavioural change – change beyond that which the pandemic has brought about? These are the key questions addressed here.

The coronavirus pandemic has exposed transport systems to a major shock that has transformed travel patterns, at least temporarily and possibly in the longer term. The initial lockdown substantially reduced our access to people and places, lessening our opportunities and choices. Yet there were benefits, too: tailpipe pollutants were down, carbon emissions fell, noise moderated, urban air quality improved and traffic congestion diminished. But subsequently people felt safer while socially distanced in their private cars, meaning that traffic returned and the benefits of reduced traffic were lost.

Sharing space with others on crowded buses, trains and planes was always unappealing, but the pandemic increased this sense of discomfort, despite the efforts of the operators to decontaminate their vehicles and limit the number of people travelling. Some people coped by switching to active travel – walking and cycling – and many more by staying at home, whether working from home, shopping and meeting via the internet instead of in person, or viewing the ever-widening range of films and performances that are available online at modest cost. Overseas holidays were replaced by 'staycations' at places accessible by car.

As widespread vaccination was achieved and we emerged gradually from social distancing and other measures intended to reduce transmission of infection, we began to see whether long-term changes in travel behaviour were likely. At the most desirable end of the spectrum would be changes that reduce the contribution of travel to climate change. Also very beneficial would be changes that ease the frustrating consequences of dense road traffic and peak-period use of public transport. The need to travel is reduced if we continue to do more domestically and are more flexible in the timing of trips beyond the home. The shock of the pandemic has shown us that there are alternatives to previous travel patterns, and indeed to travel itself. The question is whether the alternatives are superior, so that we do not revert to past behaviour. Will we see breaks in past trends of travel-demand growth?

The World War II poster in figure 0.1 was intended to discourage rail journeys by the general public at a time when the rail network was under great pressure. It was coping with bomb damage, with constraints on coal supplies, and with high demand to move troops, both as formations to new locations and as individuals travelling on leave. The same question was asked by Network Rail at the height of the pandemic, prior to Easter 2020: 'Is your journey necessary? Please only travel if it is essential.' This question, asked at times of temporary stress on the transport system, needs to also be posed in relation to our present and continuing concerns about sustainability, and particularly the contribution of the transport system to climate change.

To address these issues, I first outline (in chapter 1) the transport system that we currently have, how we use it and the problems that arise. I focus on Britain, partly because I best understand the place where I live, and partly because UK transport statistics are world-leading in both range and detail (with credit due to the Department for Transport's statisticians). I discuss travel and transport within London, in particular, where again we have access to world-leading travel statistics (credit here to the compilers of Transport for London's annual *Travel in London* reports). But I also draw on the experiences of other developed

Figure 0.1. World War II poster. (*Source*: National Archives.)

countries, in order to identify both common features and contrasting experiences.

In chapter 2 I reflect on how our transport system has developed over time, particularly over the past two centuries, since the earliest railways (drawing on my previous book in this series: *Travel Fast or Smart?*). A succession of technological developments allowed us to travel faster, which we used to travel further, constrained by the limited time we have to be on the move within the 24-hour day.

Looking back to understand how changes in technology have affected our travel behaviour is helpful when we come to look forward to assess the likely impact of further investment in established technologies (in chapter 3) and in new technologies (in chapter 4). This route offers promise when it comes to meeting our need for mobility while countering the harms that arise from motion faster than we can manage under our own muscle power.

Having set the scene with an account of where we were, prior to the pandemic, and how we had got there, chapter 5 turns to the impact of Covid-19 up to the completion of drafting this book (February 2022). I also look into the future to assess the likely 'new normal': how we will live our daily lives and what that means for how we get around. In chapter 6 I ask how we might create a better world for travel in the light of the pandemic experience and the technological options available, and then in chapter 7 I focus on the specific challenge of decarbonising the transport system, including whether we will need to reduce the amount of travel we undertake.

This book summarises a vast amount of available information about how and why we travel and about the transport system that has developed to meet our need for mobility. Some of this information takes the form of data, which I display as charts in order to show trends as well as breaks in those trends. But much of the extensive evidence and theory that is relevant to why and how we travel is found in papers in research journals. 'Transport studies' as an academic area has expanded substantially in recent years, as judged by the numbers of papers in the peer-reviewed journals. As a consequence, it can be hard to see the wood for the trees. Yet much of the output of transport researchers is increasingly divorced from the needs of practitioners – those responsible for planning and operating transport systems. For example, a substantial research literature has been created on the likely impact of autonomous vehicles, but this theoretical modelling is not based on empirical observation because there is as yet hardly any real-world experience of driverless cars. This means that the conclusions vary hugely, depending on the assumptions made about the future performance of such vehicles.

In contrast, many of us involved in transport research have seen ourselves as being based in disciplines such as engineering, economics, planning and the environmental sciences. The purpose of our research has been to advance understanding and thereby contribute to practical solutions to the problems of the transport sector. We have not seen ourselves as being primarily involved in developing a branch of knowledge through academic scholarship.

I came to academic research on travel and transport relatively late in my career, after two decades of policy work in the UK government's energy and transport departments and subsequently leading a project aimed at fostering research on ageing and its implications. Accordingly, my focus has been on how travel behaviour is influenced by key determinants including technology, economics, demographics, and policy and investment. I ask first 'What's going on here?' as a preliminary to the second question: 'What can we do about it?' My approach is analysis before advocacy. I try to stand back from theoretical detail, in order to see the big picture. I have come to the conclusion that much of the theory underpinning orthodox transport economics and modelling is problematic, and therefore misleading for investment and policy decisions – something I will explain in the course of this book.

I tell a story based on the most cogent evidence we have about why and how we travel, and about how this may change in the future. I amplify the narrative with charts that show changes over time – not the usual trend lines showing continuous growth, but the more interesting absence of growth or breaks in trend signalling a development of interest. A recent book in this series – *Transport for Humans: Are We Nearly There Yet?* by Pete Dyson and Rory Sutherland – focuses on the contribution of the behavioural sciences and complements this book. My arguments here will, I hope, be accessible to those who have an interest in how society and the economy may develop, particularly in an era in which tackling climate change is a policy priority. I also hope that professionals and practitioners in the field of travel and transport may find my analysis of use. Although the book is relatively concise, I have summarised the key ideas in a short final chapter.

Acknowledgements

I am grateful for helpful comments from Diane Coyle, the series editor, and from Richard Baggaley and Sam Clark of LPP. Thanks are also due to the Centre for Transport Studies at University College London for hospitality, and to the Centre's director, Nick Tyler.

Chapter 1

From where do we start?

The development of the modern transport system over the past two centuries has transformed our lives and vastly expanded our horizons by enabling us to travel faster than walking pace. We now have access to a huge range of people, places, opportunities and choices – whether of homes, jobs, shops, schools or other services – as well as long-distance destinations that were not previously practical. Yet this step change in mobility has come at a cost: urban air pollution, a large contribution to global warming, road traffic congestion, crowding on trains and buses at peak times, deaths and injuries from vehicle crashes, conflicts between cars and people for urban space, traffic and aircraft noise, and the concreting over of the countryside. We have tried to tackle these problems but generally have had limited success.

This chapter describes how we benefit from the transport system we have, as well as discussing the problems we have inherited, particularly those arising from the popularity of the car. This is the starting point for an exploration of how we can remedy the problems while improving the travel experience.

Why and how we travel

Each year the UK Department for Transport commissions a National Travel Survey (NTS). This covers personal travel within

Great Britain, by residents of private households in England, along the public highway, by rail or by air. (Scotland and Wales carry out their own surveys). Travel off-road, or for commercial purposes, is not included, nor is international travel by air. Some 14,000 individuals, chosen to reflect the population as a whole, complete seven-day travel diaries recording all their journeys.

The 2019 NTS report is the most recent prior to the pandemic. Averaged across the population, we made 953 trips per person per year, taking 370 hours in total (very close to an hour a day) and covering a total of 6,500 miles. The average journey length was 6.8 miles. Most of these journeys are for daily travel – the activities that get us out of the house each morning – but also included are less frequent longer trips, e.g. those for holidays.

The car is the dominant mode of travel, responsible for 61% of all trips and 77% of total distance travelled, counting the travel of both drivers and their passengers. Seventy-six per cent of households own at least one car. Eighty per cent of men aged 17 and over and 71% of women hold a driving licence. This major role for the car meets our need for convenient mobility. A recent survey found that 87% of car owners agreed, strongly or slightly, that their current lifestyle required ownership of a car, and 95% agreed that they enjoyed the freedom and independence their car ownership gave them.[1] Another analysis of survey data found that 69% of the population had personal access to a car that they could drive whenever they wished, and 87% used a car as either driver or passenger at least once a week. The same analysis found that personal car use was important for accessing employment, services and social participation.[2] In towns, as opposed to cities that have better public transport, around 80% of commuting to work is by private transport.[3]

The popularity of the car surged in the second half of the twentieth century. In the 1950s, public transport was the more important travel mode, but the growth of car use led to reduced use of buses in particular, and hence to the decline in bus services: a self-reinforcing process. Yet the dominance of the car is also the source of most of the current problems with the transport system.

Compared with the car, the other fast modes of travel comprise a much smaller share of journeys nationally: buses are responsible for 5% of trips and 4% of distance; rail for only 2% of trips but 10% of distance. Of what are known as the 'active modes' (previously called the 'slow modes'), walking is responsible for 26% of trips but only 3% of distance; and cycling for 2% of trips and 1% of distance.

The most common journey purpose was shopping, comprising 19% of all trips, followed by commuting (15%), visiting friends (14%) and accessing education (13%). Business travel was responsible for only 3% of all journeys, but 9% of distance. Leisure trips of all kinds amounted to 26% of the total. Travel patterns vary with age: people in mid-life (age range 40–50) travel most (18% more than the average); children and young people under the age of 20 travel less (12% below average); and those aged over 70 travel least (16% less than average). Journey length varies with purpose, commuting trips being the longest at 31 minutes on average, and escorting children to school the shortest at 14 minutes.

These figures are a snapshot at one point in time, immediately before the pandemic hit; I will discuss trends over time in the next chapter. The figures are also averages for groups within the population. Within each group, there is a range of travel behaviour, from those that rarely leave home on account of disability to the super commuters who travel considerable distances each day. The range reflects, in part, inequalities in society – of income, education and health. I will return later to inequalities in relation to transport.

There are significant differences in travel behaviour depending on the geography of where one lives, particularly for car use, which is responsible for 48% of trips in urban conurbations as opposed to 83% in rural villages and beyond. In London, car use is especially low – only 35% of trips – which is partly a result of the extensive public transport system, and partly due to traffic congestion and limited parking space. Car ownership runs in parallel: only 5% of households in rural villages own no car, compared with 45% of households in London.

Beyond Britain, use of the different modes of travel varies quite widely, depending on both geography and history. North American cities that grew in the era of the automobile have high levels of car use while compact European cities such as Amsterdam and Copenhagen are famous for having high numbers of cyclists. There will be more on the range of international experience later.

The car in society

Problems of popularity

The modern motorcar has many advantages for travel over short-to-moderate distances, at least when road traffic congestion is not excessive and provided parking is available at both ends of the journey. Door-to-door travel in your private conveyance is then speedy and convenient, and feels inexpensive if the purchase cost of the vehicle and annual insurance and maintenance costs are put out of mind (something that is easy to do). Cars can carry passengers, shopping and other stuff, and at no appreciable extra cost. You can keep regularly used gear in the boot and child seats in the interior. Journeys with more than one destination are readily achievable. Physical exercise is not required, so no sweat (but, regrettably, no healthy exercise). And there are evidently feel-good factors at play when people choose the cars they acquire from the huge range of models on offer: see, for instance, the growing popularity of chunky Sports Utility Vehicles (SUVs), the off-road potential of which is rarely tested. (Where are the sleek, low-slung sportscars of yesteryear?) That people are willing to pay substantial sums to own cars that are generally parked for 95% of the time attests to their perceived value.

And yet the disadvantages of the mass-market motorcar are many. The most immediate is road traffic congestion. This mainly arises in areas of high population density where car ownership is also high. There is insufficient road capacity to accommodate all the car journeys that might be made, so the traffic becomes congested and delays increase. For most of us, time is in short supply, given the many activities we have to fit into the 24-hour day. Some potential car users are deterred by the prospect of delays and make alternative choices: a less congested route or a quieter time to travel; a different mode of transport where there are viable options, as there often are in cities; a different destination, e.g. shopping in one place rather than another; or not to travel at all – working at home, for example, or shopping online.

Congestion tends to be self-limiting. If the volume of traffic builds, delays increase and more potential users are deterred. Traffic grinding to a halt is not an everyday occurrence. Gridlock occurs when some unanticipated event takes place, such as a pile-up on a motorway. This self-regulating characteristic also means that relieving congestion is difficult. If we add to road capacity by road widening or new construction, the existing traffic flows faster, delays are reduced, and drivers who were previously deterred are attracted back into their cars. The additional traffic restores congestion to what it had been, hence the maxim that we cannot build our way out of congestion, which we know from experience to be generally true.

The same argument applies to other ideas for mitigating road traffic congestion, such as encouraging modes of travel other than the car, consolidating freight into fewer vans, or even road pricing at levels likely to be publicly acceptable. Removing some motorised road users in these ways in effect creates space for others to take their place. I will discuss these options and others in more detail later. But to anticipate my conclusion, congestion is, in practice, likely to be a general feature of road travel in all densely populated, prosperous areas.

A similar conclusion applies to crowding on buses and trains, which is a problem when travelling in or between areas of high population density where travel demand at times of peak usage exceeds the comfortable capacity of the vehicles. To some extent, demand can be managed by pricing, with the highest rail fares being charged during the peak of demand, and with off-peak reductions being offered. For air travel, where all passengers must be seated, demand-responsive fares maximise occupancy and revenues for the airlines. But for commuting into and within busy cities, fare flexibility seems unlikely to make much impact on the crowding experience.

Harms to health

Road traffic congestion is a problem with which we can live. It is inconvenient but is not itself harmful. The most direct harmful consequence of mass car use is poor urban air quality. The original concern in this regard was ozone emissions from petrol engine tailpipes, which was effectively tackled by inserting the catalytic converter into the exhaust system. The main problem then became emissions of oxides of nitrogen (abbreviated NOx) and fine carbon particles from diesel engines, both of which damage health. The scale of this problem was inadvertently boosted by government policies to encourage the purchase of diesel-engine vehicles, which emit less carbon dioxide (CO_2) than their petrol-engine equivalents but more fine particles. And the problem of NOx emissions was further compounded by a substantial discrepancy between the performance of diesel engines under laboratory testing conditions and on the road: a vehicle that complied with regulations for emission levels in the lab could emit well in excess of the limit when on the road. This discrepancy encouraged car manufacturers to tune their engines to meet lab testing requirements – and even, in the case of Volkswagen, to cheat – having no regard to the consequences for urban air quality.[4]

Air pollutants from road vehicles are of particular concern as risk factors for cardiovascular and respiratory disorders. A much-quoted estimate is that some 40,000 deaths per year in Britain could be attributed to NOx and particulates, through making existing illnesses worse and bringing forward deaths by an average of seven months each.[5] This estimate was based on the advice of a government-appointed expert committee, whose published papers discuss the difficulty of attributing health consequences to the different kinds of pollutant and their different sources – information that is important for decisions on the most cost-effective interventions to improve public health.[6]

Action by environmental campaigning groups drew attention to these health concerns and technology shortcomings, resulting in many remedial actions being taken. New laboratory tests were introduced that better reflected real-world driving. Developments in engine technology improved the emission performance of diesel engines in particular. Clean Air Zones to limit the use of polluting vehicles have been implemented in urban areas of Britain where emissions exceeded permitted levels. London has had the highest concentration of NOx and has introduced the largest and most stringent clean air zone, known as the Ultra Low Emission Zone, where the most polluting vehicles are charged a daily fee (currently £12.50 for cars). The scheme was initiated in 2019 within the central London Congestion Charge zone, where, after ten months of operation, NOx levels at road side sites were down by 44% compared with a year before and 79% of vehicles were compliant, meaning that they were not required to pay the charge.[7] The ULEZ was extended in October 2021 to cover the whole area within the North and South Circular Roads.

Technological improvements that have substantially reduced tailpipe emissions harmful to health have drawn attention to other sources of harm from vehicles, particularly the fine particles released from the wear of tyres, brakes and road surfaces.

There is scope for better technology to reduce release from these sources.[8] In the past, the priorities have been to maximise durability and performance, and to minimise noise, and now we must add minimising the release of fine particles. Electric vehicles employ regenerative braking, capturing vehicle energy to charge the battery rather than dissipating it as heat, which lessens the need for frictional braking. Fine particles are also generated on the railway from braking and from friction between steel wheel and steel rail, although the evidence indicates that the coarser iron oxide particles that are produced are unlikely to represent a significant risk to the health of workers or commuters on the London Underground.[9]

The impetus for tackling road transport's contribution to poor urban air quality has come from environmental campaigners persuading responsive politicians, with the motor manufacturers in the doghouse. Air quality researchers have provided the evidence of harm to health, and the European Union has set legally binding standards based on recommendations from World Health Organization experts. This legal requirement has enabled environmental lawyers to take the UK government to court and win cases where they have alleged that insufficient effort had been made to tackle excessive air pollutant levels.

Car crashes

This legal approach to remedying poor air quality contrasts with how we attempt to reduce deaths and injuries from road traffic crashes (also known as accidents or incidents). In Britain some 1,750 people are killed each year on the roads, and more than 25,000 are seriously injured. This does not prompt public outcry for two main reasons: we have become habituated to this scale of carnage, which nevertheless is relatively low by international standards; and deaths mostly occur singly, and rarely to ourselves or those we know (we are much more sensitive to multiple fatalities, whether they are on the roads or railways or in the air). Were the motorcar to be a new invention, its use would surely be limited to highly trained professional chauffeurs rather than being available to all us amateur drivers.

One way of reducing deaths from traffic is by investing in safer roads. Modern roads that are purpose-built for cars, notably motorways, are safer than historic routes that wind their way through the countryside. The economic case for road investment takes account of the benefits from casualty reduction. The value of a death avoided is put at about £1.5 million. While there is debate about both how this figure is derived and the precise value, the point is that the benefits of fewer fatal and serious accidents are an integral part of the estimation of the expected costs and benefits (that is, of cost–benefit analysis) of a transport investment.

The economic rationale for taking account of the safety benefits of road investments may, nevertheless, be disputed in particular cases. One element of current investment in national roads is the so-called Smart Motorways Programme. This involves converting the hard shoulder of a motorway, originally intended as a refuge for broken-down vehicles, to a running lane – a change justified by the improved reliability of vehicles and as a cost-effective means of increasing capacity. There is accompanying electronic messaging to control traffic speed and manage flows in the event of an incident that blocks a lane. However, public concern and media attention have been drawn to a number of fatal crashes involving stationary and moving vehicles that, it is believed, would have been avoided had a hard shoulder been available. People feel they would be safer if there were still a hard shoulder. Data analysis, though, shows that in terms of fatality rates, smart motorways are the safest roads in the country. Per mile travelled, fatal casualty rates are a third higher on conventional motorways, and more than three and a half times higher on A-roads.[10]

A number of improvements are being made to reassure the public about the safety of smart motorways, including improved detection of, and responses to, stopped vehicles, and increasing the frequency of places to stop in an emergency. Nevertheless, the members of the House of Commons Transport Committee remain unconvinced. Their report of November 2021 recommended that the government should pause the rollout of new smart motorway schemes until five years of safety and economic data are available for every functioning scheme introduced before 2020.[11] In January 2021, the government accepted this recommendation.

Some countries, regions and cities have adopted 'Vision Zero': the principle that it can never be ethically acceptable that people are killed or seriously injured when moving within the road transport system. In effect, this puts an unspecified, but very large, monetary value on a life. Yet this concept provides no guidance as to what expenditure would be justified to reduce casualties, given all the other demands on public expenditure. Moreover, Vision Zero is a principle without legal backing that would enable litigation to enforce compliance. Nevertheless, it is reported that both Oslo and Helsinki have achieved zero pedestrian deaths through reducing speed limits, improving street design and deterring car use.[12]

Tackling harms from crashes is achieved by including estimated monetary costs for deaths and injuries avoided in the cost–benefit analysis of proposed investments. This contrasts with the approach to controlling the emission of noxious air pollutants from vehicles, which is based on legal limits. The latter could result in what some might regard as excessive expenditure on mitigation since the standards set are based on judgements about the medical effects of pollutants. Such judgements are difficult, in part due to the problem of attributing the contributions to morbidity and mortality of the different sources of air pollution, from transport and from many other origins. There is also the difficulty of eliminating confounding factors, e.g. that people on low incomes may be more likely to suffer from poor health for a variety of reasons, including living in more polluted neighbourhoods. Moreover, researchers in this area naturally tend to emphasise the significance of their findings of harms, which may be quite small in respect of risk to individuals but which, when scaled up to susceptible populations, may have significant public health impacts. Given the inevitable uncertainties, those making judgements about health impacts would naturally tend to err on the safe side and recommend low limits for permitted pollutant concentrations.

The main solution to the problem of poor urban air quality from road transport will be the electrification of vehicles, which is being driven by concerns about climate change (as discussed below). As the share of vehicles that have electric propulsion increases, there will be a decreasing need for Clean Air Zones that prohibit certain vehicles from entering or charge those that do. Nevertheless, the legal limits for air pollutants require such zones to be put in place, probably for relatively few years of operation. In contrast, decisions based on economic analysis might conclude that installation of a short-term Clean Air Zone did not represent good value for money in dealing with a problem that was on its way to a solution by other means, and that the expenditure could be better employed elsewhere.

There is no right or wrong as to whether decisions to spend public money to lessen harms are better taken within a legal or economic framework. The economic approach allows an explicit trade-off between all costs and benefits – or at least those to which a monetary value can be attributed – whereas the legal route clarifies achievement. In an ideal world, the two approaches would be complementary – blended together to specify concrete outcomes achievable through the most cost-effective means. This thought is particularly relevant to the next topic: climate change.

The least immediate adverse impact of the car is its contribution to global warming. Climate change is probably the most difficult challenge faced by society, in that so many aspects of our lives are dependent on the availability of low-cost fossil fuels. Transport is responsible for 28% of Britain's greenhouse gas emissions, with road transport contributing the bulk of this. The approach adopted in Britain to tackling our contribution to global warming has been to enact legislation requiring the country to achieve net zero greenhouse gas emissions by 2050. This means that any remaining emissions – most likely from aviation – must be offset by the removal of equivalent amounts of greenhouse gases from the atmosphere, e.g. by carbon capture and storage in deep geological reservoirs. But to achieve the Net Zero objective, there are many options of both technology and behaviour that are being identified and progressed. I will discuss later the ways in which the transport system could be decarbonised, in terms of both the feasibility of the deployment of the necessary measures and the timescale for introducing them.

Cars interacting with other road users

One problematic feature of car-based mass mobility is the conflicts that arise between drivers and other road users – pedestrians and cyclists – although of course each of us can play two or all three of these roles at different times. A principal cause of road traffic casualties from crashes is that most roads serve all classes of user. Motorways from which non-motorised vehicles are excluded are much safer than roads that accommodate mixed traffic. Segregating the different classes of road user on town roads can also be beneficial. For instance, creating cycle lanes that are separated from general traffic by raised kerbs encourages cycling by those who feel anxious in mixed traffic. On the other hand, segregating pedestrians from traffic by means of physical barriers is nowadays seen as detrimental to the sense of place that makes a town or city centre attractive.

More generally, city streets have dual functions: movement and place. For each street there is a balance to be struck between the movement of people and goods, and as a place where people live, shop and engage in a wide variety of social and economic interactions. In the early days of growing car ownership, from the middle of the last century, there was a general move to adapt towns and cities to accommodate more vehicles, both on the move and parked. As the impact of traffic on the urban environment increased, there was a reaction from people and their politicians in many historic cities, leading to the car being pushed back to create more road space for buses, bicycles and walking. Pedestrianisation of substantial areas of city centre has generally been controversial when proposed but rewarding when implemented. Shop owners tend to be anxious about possible loss of business if customers cannot park nearby, but they usually find that removal of traffic makes the area more attractive, leading to increased footfall. Nevertheless, proposals to prohibit motorised traffic are the subject of continuing controversy.

Control of kerbside parking has long been employed to optimise the use of limited urban road capacity. This has the higher purpose of limiting car use in medium-sized and larger cities, given the need to be able to park at destinations if the car is to be the mode of choice. Experience suggests that parking constraints are the most direct, and probably the most effective, way of reducing car ownership in cities.[13] However, restricting parking can be unpopular. People complain that local authorities increase charges to raise revenue, which can indeed be a significant source of funds, albeit that those funds must be used for transport projects.

The most extreme form of parking control is to forbid it altogether, which is sometimes the case for newly built apartments in inner London boroughs where public transport provision is good. This does not appear to diminish the attraction of such properties.

The most extreme form of the cars versus people dilemma arises where roads with large volumes of traffic cause severance of neighbourhoods – a problem both in cities and beyond. Some of the more substantial urban one-way and gyratory systems have been reverted to two-way operations, and this has proved popular, e.g. the closure of the north side of Trafalgar Square to traffic, allowing direct pedestrian access to the National Gallery. In the countryside, the noise from traffic on major roads detracts from the tranquillity we value. Proposals for new or widened roads are resisted, being seen as concreting over the countryside, but there is general acquiescence regarding existing roads, to which those living nearby have become accustomed.

Air travel

Air travel is used relatively little in Britain for day-to-day travel, but it is important for trips abroad, both for leisure and business. Air travel generates some of the same adverse impacts that arise from car use for surface travel. Aircraft noise can be intrusive if you live on a flight path to a major airport. Greenhouse gas emissions from aviation are of increasing importance and will loom ever larger as the electrification of surface travel proceeds. The expected continual growth in air travel prompts plans to increase airport capacity by adding runways, which are controversial on account of the additional emissions from both more aircraft and local traffic.

The availability of low-cost flying has been almost as transformational for leisure travel as the car. Nearly 300 million passengers passed through British airports in 2019. Most were on leisure trips: 46% of the total were UK residents going abroad, 28% were foreign residents visiting Britain, and 6% were leisure trips within the UK. Only 19% of all passengers travelling were on business.

Much of the growth in leisure travel has arisen because of the very efficient operations of the low-cost budget airlines, which flex their fares so as to fill nearly every seat, thereby maximising revenues. Low-cost flights out of peak season have created a market for city breaks. A survey of flying by those aged 20–45 in 2019 found that half the flights by men were for stag parties and a third of the flights by women were for hen parties.[14] However, half of UK residents do not fly at all in any given year; a quarter make one flight abroad while 8% fly four or more times. These finding are relevant when I later discuss the ways in which greenhouse gas emissions from aviation might be reduced.

Public and/or private

A feature of Britain's transport system is its mix of public and private operations – something that has changed over time and that continues to evolve. Conceptually, public sector operation offers the possibility of integration across the transport modes, while private sector involvement allows competition that improves efficiency, fosters innovation and is more responsive to customer needs. However, the optimal balance between public and private is far from clear.

Roads in Britain are publicly owned and funded out of taxation, for both maintenance and new investment. One exception is the 27-mile-long M6 toll road that was built with private finance and opened in 2003 (an experiment that has not been repeated). In other countries, toll roads for which users are charged are common. For publicly financed roads, the justification for investment in additional capacity is a central concern of the discipline of transport economics (which will be discussed later).

Railways in Britain were built and operated in the nineteenth century with private capital. They were taken into public ownership after World War II and then largely privatised in the 1990s, with private sector companies operating the trains and a public sector body managing the track, electric power and signals infrastructure. There has been very little day-to-day competition possible between operators on any particular route; rather, companies have competed for operating franchises that last seven years or more. Yet such competition, which generally involves making a payment to the government for the privilege of collecting fares from passengers, tends to encourage optimism in forecasts of demand. If demand falls short of inflated expectations, the operating company makes a loss and may have to abandon the franchise, as has happened on a number of occasions. This has deterred competitive bidding by new entrants. On top of this, the coronavirus pandemic wrecked the finances of the rail operators, which had to be bailed out by the government.

The outcome was a new policy, announced in May 2021, whereby the fragmented industry of the recent past would be replaced by a new entity, known as Great British Railways, that will be responsible for both infrastructure and operations, receiving the revenue and setting the timetable and fares. Great British Railways will contract with private companies to operate trains to the timetable and fares it specifies, in a way similar to that used by Transport for London on its successful Overground and bus networks (see below). Operators will compete for the contracts, with the aim of improving value for money and quality of customer service. Generally, the new policy should be an improvement when it is implemented, because it should lessen the inevitable conflicts between the infrastructure owner and the train operators, and particularly those over the attribution of responsibility when something goes wrong.

Motivated by a belief in the virtues of competitive markets, the political impetus that saw rail operations privatised had previously seen bus operations outside London restructured in the 1980s. Private operators were able to enter the market with minimal regulatory control and then to compete with each other for passengers on the same routes. However, it turned out that such competition was not generally profitable, and the big operators effectively carved up the routes between them, temporarily cutting fares if necessary to discourage new entrants to the market. Dissatisfaction with such outcomes led to legislation that permitted local authorities to take control of the buses in their territory, but the financial risk of doing so could be substantial. The pandemic has again necessitated a temporary government bailout here.

Prime Minister Boris Johnson, who was previously mayor of London for eight years, introduced a new national bus strategy for England in March 2021 with the words: 'I love buses, and I have never quite understood why so few governments before mine have felt the same way.' The new policy in effect reversed the approach of the Thatcher government, in which the imperative was to compete. In future, cities will be able to take control of their bus services in the way that has been successful in London, operating a more integrated public transport system with the benefits that that brings – things like simple ticketing payable with a contactless card.

Buses in London – part of an integrated public transport provision – have not been subject to the competitive regime enforced elsewhere in Britain. Rather, the operations of each bus route are from time to time put out to tender by Transport for London: the public body responsible for all public transport in the capital plus the city's major roads. Private operators compete for the routes in return for a management fee that reflects performance, accepting no risk of fluctuations in revenue arising from changes in the economy. Parts of the London rail network – the Docklands Light Railway and the Overground – are also operated by private companies, but the Underground routes are managed by Transport for London itself. Generally, public transport in London ranks highly in international comparisons and provides a good model for other British cities. Yet, inevitably, the pandemic has necessitated an infusion of substantial funds from central government.

Transport for London reports to the mayor and the responsibilities of both cover the same area, although the thirty-two boroughs and the City financial district have responsibilities for local roads and parking, which makes consistent provision for car clubs (on-street hourly car rental) across the city difficult to achieve. Beyond London, governance of transport can be quite complicated. For instance, for Cambridge and the surrounding area, there is a city council, a district council, a county council, the Greater Cambridge Partnership and a mayoral combined authority, all with different responsibilities for transport, some of which can be conflicting. More generally, transport policy and investment cannot be pursued in isolation from wider concerns about housing provision, business location and the environment, which means that governance arrangements are important.

There are a number of regulatory bodies that are charged with oversight of components of the transport sector. The most effective is the Civil Aviation Authority, which regulates safety, security and consumer protection at airports (which – apart from a handful owned by local authorities – are in the private sector in Britain) and for airlines using UK airports. The Office of Rail and Road is effective at overseeing rail safety but much less effectual in its other functions. Local authorities have some power to regulate taxis in their areas, but taxi regulation has failed to adapt to the ride-hailing revolution (notably, the entry of Uber into the market). More generally, the regulatory framework does not encompass innovations such as e-bikes, e-scooters and other cycles for hire, or Mobility as a Service, although some initial efforts by the government to make progress are underway.[15]

Generally, it would be hard to argue that the transport system in Britain is well organised and governed. To achieve better outcomes, changes are needed, as I will discuss later.

Inequalities

The increase in inequalities within society is a concern that has emerged in many developed countries since the beginning of the twenty-first century, and particularly following the 2008 financial crisis. The problem is seen in the widening range of incomes and wealth, which enables the better-off to get access to superior housing and services of all kinds. However, transport is a relatively egalitarian domain, in that there is limited scope for travelling faster by paying more. Travelling first class on trains and planes gets you to your destination no sooner. A Porsche Cayenne with an advertised top speed of 150 miles per hour makes no faster progress through congested traffic than my 2004-registered 1.2 litre Fiat Punto, and it is subject to the same speed limits on open roads. For the very rich there are helicopters and private aircraft, but the numbers using them are very small. The cost of travel is a consideration, of course. Being able to afford a car is important if that is the only practicable means of travelling to work. Some public transport users are subsidised – most importantly those on state pensions, who get passes for free local bus travel. The original rationale for this measure was that pensioners had low incomes, but that is no longer generally the case (although some 2 million households in the UK are eligible for Pension Credit to top up their income to a guaranteed minimum level). The benefit of a free bus pass depends on there being buses that serve useful destinations with reasonable frequency, which is more likely in urban areas than rural ones. There is therefore a question as to whether free local bus travel for all pensioners is good value for public money, given the many other demands in the transport sector and beyond. Present arrangements cost around £1 billion a year, yielding benefits to eligible bus users estimated to be worth a similar amount. This therefore equates to a benefit-to-cost ratio of about one, which is poor value for money compared with other uses to which the funds could be put.[16]

Aside from income inequalities, people with disabilities can find parts of the transport system difficult to navigate, particularly historic railways. There has been considerable progress in ensuring that new vehicles and infrastructure are designed inclusively, avoiding obligatory steps both on entry to vehicles and within stations and airports. It is far more cost-effective to build inclusivity into new design that to adapt existing structures.

There can be conflict between policies that recognise – and even attempt to ameliorate – income inequalities and other policy objectives, particularly environmental ones. Tax on petrol and diesel fuel in Britain has remained unchanged for a decade, reflecting the unpopularity of any increase, particularly for low-income motorists, even though there is a case for increasing the rates of duty to encourage more efficient vehicles and dampen sales of gas-guzzling SUVs. A proposition to introduce a road user charge would need to be subject to similar considerations of equity.

While the transport sector is not at the heart of the debate on income inequalities, the importance of travel to people's quality of life means that opportunities should be taken to mitigate travel inequalities at reasonable cost. However, transport differs from services such as health and school education, where public provision funded from taxation can aspire to meet the needs of the whole population, subject to ability to benefit. Beyond income inequalities, there are differences in travel patterns according to age (mentioned above), gender and ethnicity, although these do not seem to be central to the problems of the transport system discussed in this book.

In this first chapter I have outlined why we travel and how we travel, as of 2019, prior to the shock of the coronavirus pandemic. The car is the dominant mode in all developed economies on account of the benefits it offers those who can afford the costs of ownership and who appreciate the speed and convenience of this mode of travel. Yet high car use comes with serious disadvantages, some of which can be ameliorated through technological innovation, investment and regulation. In the next chapter, I will explain how we have got to where we are, and I will look at what past trends might imply for the future, post-pandemic. We must first answer the question, 'What's going on here?' before asking, 'What can we do about it?'

Chapter 2

What is changing?

In this chapter I explain how the modern transport system developed over the past two centuries and what this implies for the future. There were a number of breaks in trend as we transitioned from the twentieth century to the twenty-first – discontinuities that have not been generally appreciated but that are important for understanding what options we have to improve our travel experience.

The British National Travel Survey (NTS) – which was used in the last chapter to provide an account of our pre-pandemic personal travel behaviour – has been carried out for almost half a century, collecting information in recent years from a representative sample of the population on an annual basis by means of seven-day travel diaries. Figure 2.1 shows how three key measures have changed since the early 1970s. The lower trend line plots the average number of journeys made each year (also known as the trip rate), which has held steady at around 1,000 per person per year. These journeys are our daily travel: the trips we make when we leave our homes each day.

The middle trend line in figure 2.1 is the average time we spend on the move, which again has held steady over the fifty-year period at around 370 hours per person per year – close to an hour a day. In contrast, the upper line, which indicates the average distance travelled, shows a clear break in trend: growth from 4,500 miles per person per year in the early

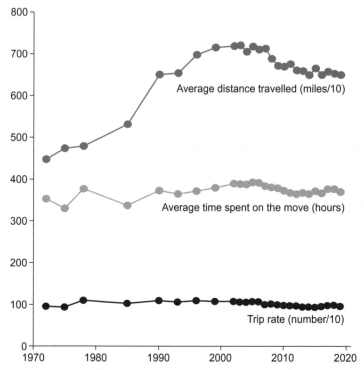

Figure 2.1. Average travel distance, time and trips per person per year. (*Source*: National Travel Survey, table 0101.)

1970s to 7,000 miles per person per year around 2000, with a subsequent fall and then stabilisation at about 6,500 miles per person per year in recent years (these data cover all modes of travel except international travel by air). Three-quarters of this distance is currently travelled by car (driver and passenger together). Car use per person has declined somewhat since 2002, as shown in figure 2.2.

The break in trend in respect of average distance travelled is the first item of evidence that travel in the present century is different from travel in the previous one. An important question is why the growth in annual distance travelled came to an end.

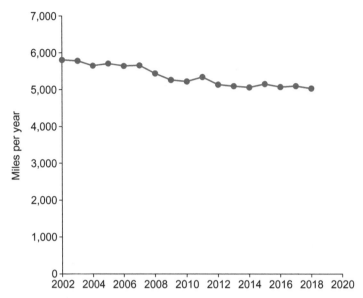

Figure 2.2. Distance travelled by car per person (driver and passenger).
(*Source*: National Travel Survey, table 0303.)

But before considering that, I need to discuss the finding that average travel time has held constant at an hour a day.

An hour a day of travel

The hour a day average travel time phenomenon is not limited to Britain. The United States Department of Transportation is responsible for the country's National Household Travel Survey, which started in 1969 and takes place at intervals of 5–8 years. Its most recent data are from 2017, when 130,000 households participated, with individuals logging travel for only one day of the week. Average travel time is not recorded, but average time spent in a vehicle as driver or passenger has been close to an hour a day since 1995. Given the dominance of the car for travel in the United States, this must be a good approximation for total travel time. Findings from similar national surveys in the Netherlands, Sweden, Denmark and New Zealand – and also from a number of studies of travel time for regions, cities and towns – indicate travel times close to an hour a day on average, not varying across a wide range of national incomes, geographies and cultures.[1]

Taken as a whole, the observational evidence is consistent with an average travel time of about an hour a day being a general characteristic of settled populations, with no consistent trend for this parameter increasing or decreasing over the years, and with no evidence of variation with national average income. On the other hand, within populations there are variations in average travel time for subgroups. Data from the NTS show that people in midlife spent some 60% more time travelling than children and older people, while people in the highest income quintile spent 70% more time on the move than those in the lowest: a reflection of more active lifestyles. In contrast, there is relatively little variation in travel time according to geographic location, except for those living in London, who spent 16% more time travelling than the national average. Similar variations with age are found in the Swedish and New Zealand surveys.

The observed invariance of average travel time has prompted suggestions about the concept of a 'travel time budget', implying that individuals have a certain amount of time they are willing to spend on travel and that they will minimise departures from that budget. However, a travel time budget is not directly observable, unlike travel time expenditure, which is measurable. The observed variation of travel time expenditure with age and income implies that individuals are making choices about activities based on the expenditure of time involved, just as they make choices about the expenditure of money. Daily expenditure of time is constrained by the 24 hours that are available to all of us, whereas money expenditure varies with income, which is unequally distributed and changes across the life course.

As with money, there is an 'opportunity cost' to time spent travelling – time that cannot be used for other desired activities – which sets an upper bound. On the other hand, reducing daily travel time tends to lessen access to opportunities that are spatially separated, as well as to intrinsic benefits of mobility that are not dependent on the destination: getting out and about, seeing the world, stretching the legs and engaging with others. The notion of a travel time budget to be attributed to individuals does not therefore seem particularly useful for practical purposes. However, the observed invariance of average travel time implies both upper and lower bounds to time that can be expended on travel, and these are considerations that can inform our understanding of travel behaviour and that should be central to decisions on transport investment and policy.

Travelling faster and further

The growth in the average distance travelled seen in figure 2.1 is necessarily the result of faster travel with unchanged travel times. This is a consequence of investment in technologies that allow us to travel at greater speeds – most recently, this has been the private purchase of more and better road vehicles and public investment in road infrastructure. Oddly, transport economists suppose that the main benefit of investments in the transport system that allow faster travel is the saving of travel time, to which a monetary value can be attributed. Accordingly, many billions of pounds have been invested over recent decades with the justification being the value of time saved – time that could be used for more productive work or worthwhile leisure activities. And yet average travel time has remained unchanged over the past fifty years, as we have seen. I will discuss the misapprehensions of the transport economists in the next chapter.

If we had the data to track backwards in time, I would expect to get to an average distance travelled of about 1,000 miles per year in the early nineteenth century. Prior to the coming of the railways, nearly everyone moved at walking pace: about three miles per hour. Within the time constraint of an hour a day of travelling, this would have amounted to about 1,000 miles per year. There were horse-drawn vehicles, but on roads that were generally poor it was mostly not possible to go much faster than walking speed, and the proportion of the population who used horse-powered transport was small.

In 1830, however, the first passenger railway in the world opened. It ran between Liverpool and Manchester, and it heralded the great boom in railway construction. There were 20,000 miles of rail routes in Britain by the end of the nineteenth century, transforming economic and social life. The higher speed made possible by the harnessing of the energy of coal allowed travellers to gain access to more desired destinations (in terms of people and places), giving people more opportunities for work and leisure and more options of all kinds.

The history of railways in Britain can be seen in passenger numbers going back to 1830: see figure 2.3. There was rapid growth in the nineteenth century, to reach a peak at the time

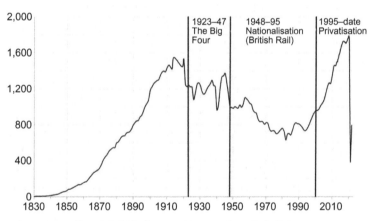

Figure 2.3. Rail passenger numbers in Britain
1830–2021. (*Source*: Wikipedia.)

of World War I, and then decline during most of the twentieth century as the car became popular. The twenty-first century has seen a rail revival, however, the start of which coincided with the privatisation of the industry. The revival may be attributed to investment in new track and trains, congestion on the roads that made rail an attractive alternative, and a shift of economic activity from manufacturing to business services, for which city centre locations accessible by rail are particularly desirable. This rail resurgence, which began in the mid-1990s, led to more than a doubling of passenger numbers by the time the pandemic hit. The rail passenger number data provide further evidence of a break in trend of travel behaviour as we transitioned from the previous century to the present one.

While the railway was good for trips between stations, the modern bicycle, invented in the late nineteenth century, increased the speed of travel for door-to-door local trips. This was important for allowing a wider choice of jobs accessible from where people lived, a wider choice of homes within an acceptable travel time from places of employment, and even a wider choice of partners in marriage. The bicycle was particularly attractive for those working in mines and factories, who could choose to live further away from dirty workplaces.

The great transport innovation of the early twentieth century was the mass-produced motorcar, which exploited the energy of oil to make possible door-to-door travel over distances greater than those comfortable for cycling – both for daily trips and for less frequent long-distance journeys. The growing popularity of the car required public investment in roads: both for improving existing routes and for creating dedicated motorways. Larger versions of the internal combustion engine that powered the car were the source of propulsion for buses and road freight vehicles. Before cars became affordable, cheaper motorised two-wheelers were popular, being faster than bicycles. Such vehicles are still very prominent on the roads of low-income countries today.

The huge impact of the car on travel in Britain can be seen in figure 2.4, which shows the share of passenger-kilometres over the past seventy years for cars, trains and buses.

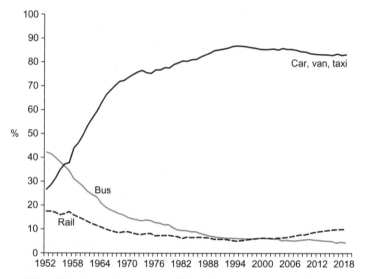

Figure 2.4. Percentage share of travel for cars, trains and buses (passenger-kilometres). (*Source*: Transport Statistics GB, table 0101.)

All these transport technologies have been subject to incremental improvements in reliability and comfort over the years. Indeed, the modern car is a marvel of reliable complexity. There have also been increases in the pace of travel, notably on the railways, where modern high-speed trains can travel at speeds of 200 miles per hour or more on dedicated track. The speed of road travel is generally limited for reasons of safety, given the variety of vehicles and the range of competencies of drivers.

There is some evidence that the hour a day devoted to travel was established when people first formed settled communities. The size of ancient settlements is consistent with an hour or so of travel being undertaken to complete daily tasks, as judged by their remaining structures, both for the area of land worked by villagers and the extent of early towns. An hour a day also seems to be the average travel time of the occupants of contemporary villages in low-income countries, where walking is the dominant mode of travel.

As new technologies that allowed faster travel arrived, so towns and cities grew in size. The narrow and often-irregular street patterns of the centres of modern historic cities reflect the original walking city. With the coming of public transport in the nineteenth century – first with horse-drawn buses and then with motorised ones, as well as trams and trains – cities expanded, as commuting from residential suburbs became feasible within the travel time constraint. The motorcar's arrival in the twentieth century allowed lower-density suburbs to develop, beyond walking distance from railway stations. For cities whose growth was predominantly in the era of the car, population density generally remained low as people chose agreeably located family homes that were accessible by car.

Looking back over time, we can therefore identify four eras of human travel. Our *Homo sapiens* ancestors walked out of Africa some 60,000 years ago to populate the whole of the habitable earth on foot. As hunter–gatherers, they were on the move for about three to four hours a day (as judged from the behaviour of the remaining societies of this type) and would cover 3,000–4,000 miles per year. With the discovery of agriculture some 12,000 years ago, people moved around less, since effort expended tending crops was rewarded by higher yields, and the hour a day of travel time was established. This second era persisted until the early nineteenth century when the invention of engines to exploit the stored energy of fossil fuels – first coal and then oil – made possible a period of increasingly faster travel, including mass travel by modern aircraft. But this third era has now come to an end, in part on account of our inability to travel faster in the time we have available, constrained by all the other activities to be fitted into the 24-hour day. The fourth era of travel, in which we are living, is characterised by the need to decarbonise the transport system, as I will now discuss.

The end of faster and further

Figure 2.1 tells us that the average annual distance travelled in Britain ceased to grow around the turn of the century. Figure 2.2 confirms that car use has also ceased to increase in recent years. Figure 2.5 shows a compilation of data for eleven developed economies in respect of car use per person per year, estimated from official statistical sources for total distance travelled by car

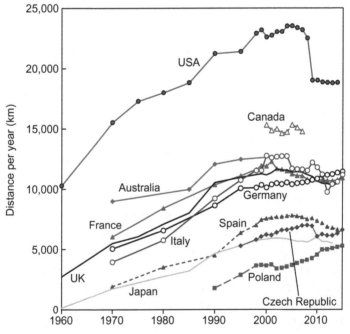

Figure 2.5. Distance travelled by car per person. (*Sources*: Dr Kit Mitchell, from UN ECE Transport Statistics, for total distance travelled by car; official population statistics for total population.)

and for total population. Car use grew until around the year 2000, at which point it stabilised or fell in all but one case (Poland).

There is good evidence, therefore, that the average distance travelled per person by car in many developed economies ceased to grow as the end of the previous century approached. This phenomenon has been termed 'peak car', in analogy with 'peak oil', which referred to the expected peaking and decline in output of this finite resource. However, for car use, the evidence points to a cessation of growth as the prime effect, with possible long-term decline not yet generally apparent. Accordingly, I propose the term 'plateau car' to designate the phenomenon.[2]

There are two main reasons why car use has stopped growing in developed economies. First, we have run out of ways of travelling faster, which means that we cannot travel further given the fixed amount of time available for travel. Second, there is reason to suppose that we already do enough daily travelling to meet our needs.

First, then, let us discuss the limitations on travelling faster. Generally, speed limits apply to all roads in nearly all countries, and their main aim is to minimise crashes and mitigate their consequences. A well-known exception is the German Autobahn network, which has no legal speed limit for cars outside urban areas. (An advisory speed limit of 130 kilometres per hour applies, though, and driving faster than this can increase liability in the case of a collision.) There seems no prospect of any general relaxation of speed limits, given the current global road safety record: the World Health Organization estimates that 1.35 million people die each year on the world's roads. Indeed, there is a move to reduce speeds on urban roads to save lives, with 20 miles per hour speed limits widely adopted in Britain's cities, encouraged by campaigners with the slogan '20's Plenty for Us'.

In or near urban areas, it is road traffic congestion rather than the legal speed limit that constrains the rate of progress. Congestion arises in areas of high population density and high car ownership, where there are more potential trips that might be made by car than there is road capacity to accommodate those trips. As traffic builds up, delays increase and some drivers make other choices – whether that is choosing a different route, time, destination or mode of travel, or not to travel at all, depending on the options available. Congestion is therefore generally self-regulating, with gridlock rare, arising only as the result of some unanticipated event.

Yet this same dynamic means that it is difficult to mitigate congestion. Adding road capacity attracts back the trips that were previously suppressed on account of unacceptable delays, hence the general experience that we cannot build our way out of congestion. Other measures intended to reduce car use are similarly nugatory: promoting public transport or active travel, for example; consolidating road freight into fewer vehicles; and charging cars on the move, as with congestion charging in London. All of these interventions take some vehicles off the road thereby allowing others to take their place. I shall discuss these possible policy interventions in more detail later, as well as the impact of new transport technologies, but for present purposes, suffice to say that there is little prospect of reducing traffic congestion to an extent that would significantly increase the speed of car travel in urban areas. The mistake made by many is to suppose that there is a fixed amount of traffic – so that removing one element would relieve congestion. In reality, the dynamic balance between trips made and trips suppressed means that a congested equilibrium is inevitably re-established.

As mentioned earlier, there is scope for increasing the speed of rail travel, and indeed High Speed 2 (HS2) will do just that. HS2 – a controversial new rail route cutting travel time between London and the cities of the Midlands and the North of England – is the largest ever single transport investment in Britain. I will return to the economic justification for this expenditure of more than £100 billion later, but for the moment, we should simply recognise that rail accounts for a minority of all trips, and HS2 would be responsible for a minority of a minority, so no significant impact is to be expected on the average speed or distance travelled.

Another innovation that will speed up one mode of travel is the increasingly popular electric bicycle (e-bike), which is equipped with a small electric motor to assist the rider's muscle power. The additional power allows faster travel and greater access to desired destinations. Electric propulsion increases the range for bikes while reducing it for cars. Yet cycling accounts for only 1% or so of personal travel in Britain (in terms of passenger-kilometres), so the effect of e-bike uptake on average distance travelled will be very small.

To complete the account of the limitations of existing transport technologies, let me mention that the speed of travel of passenger aircraft has barely increased since the Boeing 707 – the first widely adopted jet airliner – originally took to the air in the 1950s. Improvements to engine performance and airframe construction have increased the range without need for refuelling, however, which has allowed non-stop, effectively faster travel over the longest routes. Concorde, the supersonic airliner, was environmentally problematic and commercially disappointing, operating for only twenty-seven years.

In summary, there is only very limited scope for existing transport technologies to permit faster travel and more access to where we want to go. I will discuss the potential of new technologies later, but to anticipate my conclusion, they too offer quite limited prospects for higher speeds.

Saturation of travel demand

There is a second reason why the average distance travelled ceased to grow as we entered the present century. The ending of a growth phase in the demand for a product or service is a feature common to all markets. A new product that offers benefits to users is initially taken up by so-called early adopters, with the more cautious following later as experience is gained. High levels of household ownership are found with many domestic appliances, such as washing machines – a characteristic of a mature market in which substantial growth has ceased and demand is said to have saturated. It is plausible that the cessation of growth in surface travel and car use reflects demand saturation for daily travel. That is, we now travel sufficiently to access a wide choice of destinations offering services and opportunities, with little need to go farther for yet more choice.[3]

The growth of average distance travelled in the last century (see figure 2.1) is a consequence of an increase in speed, given the unchanged average travel time over the period, and is largely attributable to the growth in car ownership. The observed unchanged average travel time means that the main benefit is better access to desired destinations, and the associated increase in opportunities and choices. An important feature of access is that it is subject to what economists term 'diminishing returns', as choice and opportunities increase. For instance, having one local food store in your neighbourhood is very useful; having a second gives your more choice, which is a positive thing; having a third offers a bit more choice, but this is less valuable than adding the second, which was in turn less valuable than the first.

Another characteristic of access is that it increases with the square of the speed of travel.[4] So, switching from walking at three miles per hour to cycling at ten miles per hour increases access and choice by a factor of more than ten; similarly, a car travelling at twenty miles per hour provides more than forty times as much opportunity as walking. This is the main reason for the popularity of the car.

In practice, this argument from geometry is a simplification of the reality in that the places that can be reached depend on the roads available. This is not a problem in cities with dense street layouts, but in rural areas the road system limits travel by vehicle – that said, the places you might want to get to are generally located on those roads. In the more remote areas, there may be a single road, in which case access is proportional simply to the speed of travel.

So, we have (a) access increasing at least proportionately to the speed of travel, and generally faster than that; and (b) the benefits of more access being subject to diminishing returns. This is consistent with there being demand saturation, for which some evidence exists.

There are UK data for the proportion of the population that can access important services for health, education, food shops, jobs, etc. We find that high proportions of potential users have access within reasonable travel times. For example,, 71% of people have access to their family doctor within fifteen minutes of travel time by public transport or walking, and 96% have access within thirty minutes; 87% of people are within fifteen minutes by bicycle and 98% by car. Similar high levels of access are found for other services, including employment, schools, food stores and services located in town centres.[5]

We can also estimate how much choice people have of such services, based on data collected on journey times to these destinations. For instance, the populations of a majority of English localities have access to an average of five or more GPs within a thirty-minute journey by public transport or walking, and in almost every locality, people have such a choice within fifteen minutes by car.[6] A study of access to supermarkets in Britain found that 80% of the urban population had access to three or more large stores within fifteen minutes by car, and 60% to four or more.[7] Such high levels of access and choice are consistent with the proposition that demand saturation is contributing to the cessation of growth in per capita car use. In the case of supermarkets, this has come about through two complementary developments: the growth in car ownership, and the supermarket chains building more large stores with extensive parking, particularly by taking advantage of road construction that has made land accessible for new stores on the outskirts of urban areas. However, both trends are now largely played out. Although comparable data for other developed economies are not available, it seems likely that similar saturation phenomena have been occurring.

Supermarkets are an example of a travel destination that may be termed 'replicable', in that they can be built to meet demand. Another class of destination comprises what might be designated 'status destinations', analogous to the economic concept of 'positional goods' that are desired as status symbols. These are locations of a unique or special value that are either scarce in some absolute or socially imposed manner or subject to physical capacity limitations. Examples include historic sites, waterfront properties and Premier League football stadia. Travel at higher speeds allows access to a greater range of such distinctive non-replicable locations. However, the benefits of access are offset by increased crowding or higher prices as others with similar interests take advantage of improved travel facilities. Travel demand to get to such status destinations may therefore be expected to saturate. The contrast between replicable and status destinations is exemplified by the comparison of schools in general with 'good schools'. Many parents are keen for their children to get access to the latter and may be willing to provide car transport to locations more distant than the nearest school, but the limited number of places constrains the overall demand for travel on the school run.

One particular case of demand saturation is especially relevant to the cessation of growth of average distance travelled. The proportion of households in England owning one or more cars increased steadily until the end of the last century, at which point growth ceased – at the point when three-quarters of the population owned at least one car. There has been some subsequent growth of overall car ownership as the proportion of households owning more than one car has increased, but the first car is responsible for the bulk of the mileage travelled, with other cars likely to be used largely for local trips in place of public transport.[8]

This cessation of growth in household car ownership at the end of the previous century coincided with the cessation in growth of the average distance travelled, both by car and by all modes of travel. The former is an important cause of the latter. Accordingly, cessation of growth in household car ownership is in itself both an example of demand saturation and an important contributing factor to the cessation of growth in average distance travelled.

Car ownership is not for all

The question, then, is why one-quarter of British households do not own a car. The cost of car ownership is always a factor in the decision over whether or not to own one, and this is of course more important for low-income households than for the better off. A clue to other influences is that in the inner boroughs of London, a prosperous city, 60% of households do not own a car, while for London as a whole the proportion is 45%.[9] The extensive public transport system reduces the need for car ownership, as does the high density of services and amenities. The availability of parking space, and its cost, is also a constraint.

More generally, demographic developments have had an important impact on the pattern of travel demand and how this is distributed across the various modes of transport. The movement of populations from country to city is a long-term global trend, and one that has been reinforced in developed economies in recent years by the shift from manufacturing to business services and the 'knowledge economy', which is more likely to be located in city centres. Contributing to urbanisation is the process that economists term 'agglomeration', which offers particular advantages to the business services sector. Firms that are concentrated in one geographic area benefit from learning, sharing and matching. Firms acquire new knowledge by exchanging ideas and information, both formally and informally. They share inputs via common supply chains and infrastructure. And they benefit by matching jobs to workers from a deep pool of labour with relevant skills. Agglomeration leads to urban development and population growth at higher densities, despite high land prices, rents and other costs. This growth has increased congestion on the urban road network and thereby made car use less attractive, but at the same time it has improved the economic viability of public transport.

Both the growth of urban economies and the expansion of city centre universities have attracted young people to move to vibrant cities to study, work and live. Agglomeration benefits consumers as well as businesses in that high population density leads to greater choice of amenities (shopping, cultural and social). Dating apps offer more choice of potential partners when more people live in the vicinity. The tendency to defer settling down to suburban life until past the age of thirty contributes to the popularity of city living.

Urban growth and agglomeration have contributed to a significant change in travel behaviour among young people in developed economies. A recent comprehensive review of the research evidence and survey data noted a trend since the mid-1990s for successive cohorts of young people (aged 17–29) to own and use cars less than their predecessors.[10] This contrasts with the baby boomers, born between 1946 and 1964, who led rapid, prolonged and persistent growth in car ownership and use. Factors contributing to this trend away from car use among the young include the cost of car ownership (not least high insurance charges for younger drivers), problems of parking in cities and on campuses, and the viability of alternatives such as bicycles, public transport, shared car use and smartphone apps to summon a taxi. The main causes, however, lie largely beyond the transport system and include increased participation in higher education, for which the car is not a necessary part of the lifestyle; the use of digital communications and social media; and, more generally, a delayed transition to what was traditionally seen as adulthood, i.e. commitment to a career, getting married, home ownership, starting a family, with stable employment a strong determinant of being a car driver.

An important question is how this shift away from car use by young people will affect the way they travel as they get older. Those who start to drive later drive less; for instance, for those now in their thirties in Britain, if they learnt to drive when aged 17, they now drive 10,000 miles per year on average, but if they learnt at age 30, they drive only around 6,500 miles per year.[11] It seems likely that this reduced mileage reflects greater experience of modes of transport other than the car, gained before learning to drive, as well as the fact that the latter group is more likely to have lived in places where such alternatives are viable, in particular for journeys to work. Prior to the coronavirus pandemic, the evidence concerning the driving behaviour of young people suggested that while there were many uncertainties about the travel behaviour of future cohorts, as well as about how that might change as those people get older, it was nevertheless hard to envisage realistic scenarios in which all these uncertainties combine to re-establish earlier levels of car use. This conclusion may now need to be re-examined, however, and is something we will discuss in a later chapter.

Peak car in big cities

The discussion so far has focused on per-person travel demand, but population growth can be an important determinant of the total demand for travel. In rural areas where there is sufficient space for cars, both on the road and when parked, growth in population would be expected to result in growth in car ownership, all else being equal. In urban areas, however, this may not be the case. I noted earlier the relatively low level of car ownership in London, and I now explore further the reasons for this phenomenon, which is also seen in other big cities.

The impact of urban population growth on car use is well illustrated by the experience of London, for which exceptionally extensive travel statistics are available.[12] The population of the city grew steadily from 1 million in 1800 to reach 8 million in the 1930s/1940s and then declined as people left war-damaged, poor-quality neighbourhoods to find better homes in towns and suburbs beyond the capital's boundaries. The population had fallen to 6.6 million by the early 1990s, but the tide then turned as people appreciated the attractions of city living at greater density: higher earnings, for example, and attractive amenities, both social and cultural. A boom in construction allowed the population of London to increase quite rapidly, reaching nearly 9 million by 2020, with further growth expected (prior to the pandemic) to reach nearly 11 million on a central case projection by 2050.[13]

Despite London's population having grown by more than 2 million since the 1990s, the city's road traffic volume has not increased in recent years, as seen in figure 2.6.[14] Indeed, there has been a decline over the past twenty years, most marked in Central London, less so beyond. The reason for this is the constraints imposed by road space. Plans were formulated in the 1970s for a major expansion of road capacity in London to accommodate the expected growth in car ownership. They featured plans for an elevated 'motorway box' located in the inner suburbs, to divert traffic from central London – an innermost orbital road, to complement the intermediate North and South Circular routes and the M25 motorway near the outer edge of the conurbation. One section of the planned box was built, with a link known as the Westway running westward from the centre of the city, but this provoked strong opposition to any further such construction on account of the damage to the urban environment. The only part of the box that was constructed was the eastern side, where acceptable space could be found amid the urban dereliction from wartime bomb damage and the relocation of the docks to deeper-water ports to accommodate container vessels.

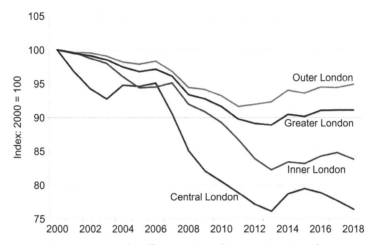

Figure 2.6. Road traffic in London. (*Source*: Transport for London's 'Travel in London, Report 12' (figure 9.2).)

London has retained its historic street pattern, which has served to limit growth in traffic. Indeed, road space for cars has been reduced as bus and cycle lanes have been constructed, and as extra space has been allocated for pedestrians, particularly in the centre of the city where there are high numbers of both residents and tourists.*

Accordingly, the population of London has been growing but traffic has not. This necessarily means that the share of journeys by car must have fallen, and indeed that is the case, as can be seen in figure 2.7. The share of car trips (driver and passenger) has fallen markedly, from 50% in the early 1990s, when the data series begins, to 35% in 2019. Active travel – walking and cycling together – has held broadly steady at about 25% of journeys, with cycling currently responsible for 2.5%. In a mirror image of

---

*An alternative view, backed with some evidence, is that use of satnav devices has resulted in an increase in traffic on London's minor roads. This has led to disagreement between Transport for London and the UK Department for Transport about trends in motor vehicle traffic in London. I will discuss this further in a later chapter when I consider the impact of digital technologies.

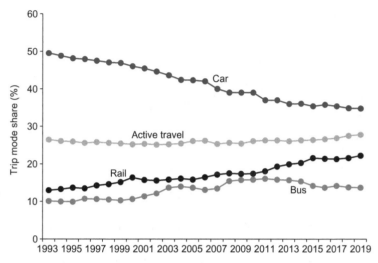

Figure 2.7. Trip mode share in London. (*Source*: Transport for London's 'Travel in London' reports.)

the decline in the share of trips by car, public transport's share has grown from 23% in the early 1990s to 36% in 2019. The rate of increase for buses slowed and then ceased from around 2010, followed by a decline in passenger numbers. A substantial increase in rail capacity is due in 2022 with the planned opening of Crossrail, a new 70-mile-long east–west route through central London.

My estimate of the share of journeys by car in London over the century from 1950, looking ahead to 2050, is shown in figure 2.8.[15] Images of streets in the 1950s show light traffic and few parked cars. Rationing of petrol, which had applied during World War II, ceased in 1950. As incomes grew, car ownership increased, and so did car use. At the same time the population of London was falling, reaching its minimum around 1990, at which point car use peaked at 50% of all trips. Thereafter, this share fell to 36% in 2019, as shown in figure 2.7. I have extrapolated to 2050 on the basis of a central projection of population growth

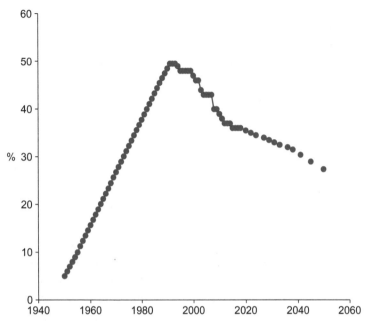

Figure 2.8. Share of journeys by car in London, 1950–2050.
(*Source*: author's estimates and Transport for London.)

(see earlier discussion) and an assumption that there will be no increase in road capacity for cars but that there will be investment in rail capacity broadly to match population growth. On this basis, the share of journeys by car would be projected to fall to 27% by 2050. The Mayor of London's 2018 Transport Strategy was more ambitious, with a target of 20% by 2041, but that was before the pandemic, which both made cars more attractive and may have affected future population growth (the implications of which I will discuss later).

The marked peak of car use shown in figure 2.8 prompts the designation 'peak car in the big city', to distinguish it from 'plateau car' (previously termed 'peak car'), referring to the average annual distance travelled per person, as I explained earlier. Here, the peak reflects the *share* of journeys by car under circumstances where car use is constrained by road traffic congestion, which in turn results from limited road capacity.

While London provides the best data sources, there is evidence of a similar phenomenon in other large cities. A recent study identified a peaking of car use in terms of car trips per day in Paris, Berlin, Vienna and Copenhagen, as well as London; on the other hand, data from US cities indicate continuing very high levels of car use.[16] There is supporting evidence of a decline in car traffic in a number of UK cities (including Manchester and Birmingham), particularly in their centres, as well as in the major Australian cities.

A broad distinction may be made between cities with historic central areas – where the street pattern limits car use and the population density makes public transport economically viable – and more recent cities built at low density with the car in mind as the main means of mobility. This distinction is broadly between European and North American cities, although central parts of some of the latter do have high population density, with the suburbs of most generally having low density. The available evidence suggests a general phenomenon whereby successful cities with well-established centres attract people to work, study, visit and live. The total population and population density therefore both increase, and authorities recognise that the road network cannot be expanded to accommodate more car use without damaging the urban environment. Decisions are consequently made to invest in public transport – particularly rail, which provides speedy and reliable travel compared with cars, buses and taxis on congested roads – as well as to improve facilities for walking and cycling. A common characteristic is the existence of a historic city centre that people find attractive for work and leisure: here, declining car use increases the attraction. In the absence of an appealing downtown district, population growth may lead to continued low-density development, with no mode shift away from car use.

In the twentieth century, increasing prosperity was associated with increasing car ownership and use in developed economies. In the twenty-first century, however, increasing prosperity is associated with *decreasing* car use in big cities that have attractive centres, growing populations and that can afford to invest in an extensive network of high-quality public transport as an alternative to the car on congested roads.

An important question for smaller cities and towns is whether they want to adopt the approach of the large, successful cities: that is, discouraging car use in order to gain the benefit of more and better economic, cultural and social interactions. The challenge is to put in place an effective, attractive and affordable transport system that can reduce dependence on the car – a topic to which I will return later.

There is a similar question for developing countries in which car ownership is still relatively low: can the peaking of car use in large, densely populated cities be avoided, with transition instead to the smaller mode share envisaged in the developed economy cities discussed above? It would make sense, logically, to anticipate a modest share of car trips in the long run, driven by the same imperatives that have applied in developed economies, but the popularity of car ownership as it becomes affordable might mean that the peak car phenomenon would be hard to avoid.

Population growth in cities that experience significant traffic congestion is best accommodated in new homes with minimal parking facilities, which are prevalent in a number of inner London boroughs. On the other hand, when an increasing population is largely housed in new homes built on greenfield sites, the occupants will generally want to use cars for their mobility needs, and new road capacity may be required.

Given that the *average* distance travelled has not increased over the past twenty years (as shown in figure 2.1), a credible central case projection of future travel demand would retain this flat trend. Accordingly, *total* travel demand will depend on the magnitude of population growth, and the pattern of that demand will depend on the location of new homes (greenfield or existing urban locations). Following Britain's exit from the European Union, official projections of population growth have been scaled back, but there is no government steer as to the balance of location of new residential properties.[17]

Little sign of 'peak air'

Travel by air has displayed a long-term growth trend. Figure 2.9 shows passenger numbers through UK airports, which increased from 44 million per year in the mid 1970s to almost 300 million in 2019, before the pandemic. A downturn was seen following the financial crisis of 2008, but growth subsequently resumed. In the case of travel between the United Kingdom and the United States, it seemed as though growth had ceased around the turn of the century, with no upward trend over the subsequent fifteen years. Significant growth resumed, however, over the three years to 2019, as can be seen in figure 2.10.

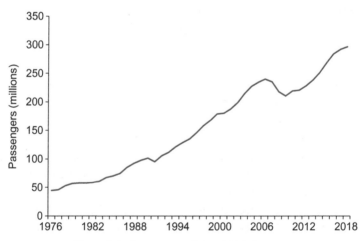

Figure 2.9. Passengers through UK airports.
(*Source*: Civil Aviation Authority.)

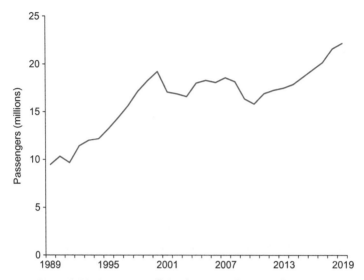

Figure 2.10. Passengers flying between the United Kingdom and the United States. (*Source*: Civil Aviation Authority.)

Travel to and from Japan from both the United Kingdom and the United States has displayed a different pattern from elsewhere. There was strong growth in passenger numbers up to around 2000, followed by a marked decline. This was due mainly to decreasing numbers of Japanese tourists, which is probably the consequence of an ageing population anxious about foreign food and making themselves understood, together with slow domestic economic growth.[18] While not typical, the example of Japan illustrates the possibility of demand saturation, also known as market maturity, in air travel.

The pandemic substantially shut down air travel. The question of resumption and future growth will be considered later.

To conclude this chapter let me summarise the breaks in trend that we observed as we moved from the twentieth century to the twenty-first.

- Average distance travelled ceased to increase, both by all modes and by car.
- Household car ownership stopped growing.
- Rail travel began a rapid rise.
- Car use, as a share of all trips, peaked in London and in other successful cities.

These breaks in trend demonstrate our capacity to respond to innovation, both by taking advantage of what is new and subsequently by deciding that what we have is sufficient.

In contrast, some aspects of travel did not change. Average travel time remained at an hour a day, and the average number of journeys at about 1,000 per year. As we will discuss later, the pandemic disrupted our usual travel patterns and disturbed these hitherto-constant features. The question is when we will return to normal, and whether the new normal will be different from the old. Before turning to those questions, though, we need to consider how the transport system we have inherited can be improved through investment in better infrastructure and vehicles. That is the topic of the next chapter.

Chapter 3

Will investment solve our travel difficulties?

The previous chapters have addressed the main problems with the transport system – road traffic congestion, crowding on public transport, poor urban air quality, deaths and injuries from crashes, traffic interposing between people, and transport's contribution to climate change – and explained why current policy approaches have failed to tackle them. Broadly, the solutions can be broken down into three types: investment in new and better facilities, changing travel behaviour, and technological innovation. In this chapter I consider the prospects for investment and changing behaviour, and then I will assess the promise of new transport technologies in the next.

Investment

The transport sector forms a substantial part of the economy and requires continual investment to maintain and upgrade both infrastructure and vehicles. Past investment has allowed the growth in the average distance travelled that was outlined earlier. Decisions about future investment will influence how we travel in the future. Investment in the transport system is generally considered to be a 'good thing', and there is plenty of it. Yet the detrimental aspects seem to persist, so we need to probe the rationale for the main kinds of investment to understand their limitations.

In what transport should the government invest?

Much transport investment involves the acquisition of vehicles, which is a matter of personal decisions by individuals and commercial decisions by transport businesses. But it is the public sector that is responsible for the road and rail networks, and here there are very large expenditures planned or already underway.

Crossrail in London will cost nearly £20 billion by the time it is completed in 2022. Construction of High Speed 2 is underway. This new rail route between London and the cities of the Midlands and the North will cost more than £100 billion, which overlaps with a regional rail investment programme worth £96 billion. A five-year programme of investment in the Strategic Road Network, comprising the motorways and other major roads, is worth £27 billion, and there is further substantial government expenditure on improvements to roads and railways that effectively takes the form of ad hoc grants from the Department for Transport (DfT). The former system of five-year investment plans for the railways proved unviable for delivering specified outcomes and has been replaced with an 'enhancements pipeline': a rolling programme of investments constrained by the availability of funding. More generally, the DfT has some fifteen different funding streams that are open to local authorities, a majority of which involve competitive bidding.[1]

Transport investment that is ultimately funded by taxpayers is subject to a test of value for money known as cost–benefit analysis. Essentially, this involves estimating the economic value of the benefits to users of the investment, deducting the value of any harms, and comparing the resulting net benefit with the cost of construction. The ratio of benefits to costs (the benefit–cost ratio, or BCR) is a measure of value. A BCR of 2 means that for every £1 spent on construction, a benefit of £2 is expected.

So what are the benefits of investment in the transport system? The main benefit that has generally been assumed is the saving of time from faster travel, which most investments aim to achieve: for instance, by widening a congested road through converting a historic single-carriageway road to dual carriageway with two lanes in each direction. The value of time saved is estimated through a market research exercise in which respondents are asked whether they prefer a journey that is faster but more costly or one that is slower and less expensive (for the same route). A representative sample of road users may be asked to express a preference between, for instance, a trip taking 4 hours and 23 minutes with a fuel cost of £33.30 and one taking 3 hours and 30 minutes at a cost of £35.00. Responses to a range of such questions allows a monetary value to be attributed to travel time.[2]

Estimating the economic benefit of a transport investment then involves multiplying the time saved per individual (usually quite small) by the number of travellers (quite large), and then by the monetary value of time. Such estimates need to be based on projections of future numbers of travellers and future travel times; the latter may increase if the demand for road travel grows, meaning that congestion increases. Future benefits are 'discounted' – that is, reduced by 3.5% (the current rate, which is subject to debate and can vary) – for each year into the future because the worth of £1 of benefit in the future is less than the same sum now, as we recognise from the need for our savings to earn interest.

Yet there is a paradox in supposing that a saving of travel time is the main benefit of transport investment, given that average travel time has remained unchanged over almost half a century – despite huge investments in the transport system over that period having been justified by the value of travel time savings. One possible explanation is that without this investment, congestion would have been worse and journey speeds would have been slower. Yet it seems implausible that the amount of investment, which has fluctuated significantly over the years, has consistently been just sufficient to avoid an increase in average travel time. Moreover, the constant hour a day of travel time applies across countries generally, which have very different investment programmes.

The better explanation is that rather than the main benefit of investment being more time for work or leisure, it is that we take advantage of faster travel to travel further in the time available, allowing us to reach more distant destinations, to engage with a wider range of people and to experience more opportunities and choices. The benefit of investment to users of the transport system takes the form of better access, not travel time savings as transport economists suppose.

Imagine that you live in a village that is inadequately served by public transport (if it has any public transport at all) and that you do not have use of a car. Your choices of shops, jobs, schools and so on are quite limited. Now suppose you acquire a car. You might initially save a few minutes travelling to the village shop, but you quickly realise that you are now able to get to the supermarket in the nearest town, which has a greater range of goods at more competitive prices. Similarly, when you come to change job or move house, car-based mobility allows you to access a much greater range of opportunities and choices, which is why there is plenty of local traffic in the more populated parts of the countryside.

The benefits of investment in the transport system are, therefore, access benefits for users. Indeed, in its November 2021 Integrated Rail Plan for the North and Midlands, the DfT at long last accepted the problematic nature of theoretical time savings:

> Over the last 50 years the time people spend travelling has remained relatively constant, though distances travelled have increased... Overall, people have taken the benefits of better transport links as the ability to access a wider range of jobs, business and leisure opportunities, rather than to reduce total time spent travelling.[3]

It is gratifying to find the DfT seemingly accepting this reality, which I have been drawing people's attention to for many years. Nevertheless, there is a footnote appended to the Integrated Rail Plan that suggests the DfT does not yet quite get it: 'Noting that the use of estimated time savings as the basis for quantifying economic impact remains robust.'

Access benefits are different in nature from time-saving benefits. As I explained earlier, access benefits tend to increase with the square of the speed of travel while being subject to diminishing returns, which means that demand for travel tends to reach saturation. Recognising that the benefits of investment take the form of improved access means that there is a limit to the value for money that can be achieved from continuing to invest. In contrast, using time savings as the measure of value implies no limit to the amount of investment, because there will always be travel time that supposedly might be saved.[4]

A problem with access benefits is that it has proved difficult to attribute monetary values to their increase. However, increased access leads to changes in how land is used and therefore to changes in the value of the land and the property located there (it is convenient to use the American term 'real estate' to designate both land and the property on it). Increases in real estate values, as seen in the market, provide a clear basis for attributing economic benefit to transport investments. According to the orthodox view, however, incorporating increases in real estate values would be double-counting benefits that have already been included in travel time savings. This implies preferring notional benefits, derived from models and not observed in reality, to actual market values – which is not, in my view, helpful in reaching decisions on investments.

To put this point in a specific context, consider a proposal to build a bypass around a village, in order to reduce the adverse impact of through traffic in a location built before the era of motorised vehicles. The economic case would be largely based on the time saved using the purpose-built road. However, a bypass may also make agricultural land accessible for new housing. Developers might seek planning consent to build on this land, which, if secured, would increase the land's value and allow new homes to be built. This would, in turn, generate local traffic, which would result in different economic benefits than had planning consent not been granted. The orthodox approach to economic appraisal supposes that the potential change in land use can be disregarded since the estimated time savings capture all the benefits, regardless of land use change (albeit that the distribution of those benefits between road users, land owners and developers would now be different). Yet in reality, the proposed bypass investment has very different economic outcomes depending on decisions made by planners and developers, and there is every reason to consider the outcomes separately. Orthodox transport economists have tried to avoid introducing the additional complexity of changes in land use in order to simplify decision making, but they have done so at the cost of departing from reality. Decisions about transport investments should not be taken within a transport silo but need to be tripartite, involving planners and developers as well.

The problems of transport modelling

To put economic analysis into practice, we need models that allow us to estimate the impact of an investment in new infrastructure or those of any other kind of intervention aimed at improving the quantity or quality of the service. Over the years, a huge range of transport models[*] have been developed and applied. The simplest models focus on road traffic and involve four stages, the first of which considers trip generation, both origins (such as homes) and destinations (such as shops), by

---

[*]These might also be designated 'travel models' since it is travel behaviour that is modified as a result of changes to the transport system.

journey purpose (such as shopping), and looks at how these might change over time. Three sequential stages aimed at optimising outcomes follow: trip distribution that matches origins with destinations; choice of mode of travel; and route assignment that allocates trips between an origin and a destination by a particular mode to a specific route. Such models can be used to compare the total travel time in a situation in which an investment is made with that where it is not, the difference being the basis for estimation of the value of the time savings.

The main limitation of this widely employed four-stage framework is its neglect of changes in how land is used. An investment that results in faster travel will make locations more accessible to more people, which increases the value of existing land and property and may make new construction attractive – which can, in turn, attract further traffic. There is a class of model known as a land use and transport interaction (LUTI) model, which includes efforts to model complex urban environments. Another limitation of the four-stage model is that it fails to recognise that when we leave home, we may visit a number of destinations before returning. An approach known as activity-based modelling aims to simulate the overall impact of the trip chains that individuals in a population might make.

The general problem with all these transport models is demonstrating their validity to the extent that their projections are accepted as sufficiently reliable to inform investment decisions. In the case of a small-scale traffic model that covers a single neighbourhood, it may be possible to test its accuracy by comparing prediction and outcome for an intervention such as changing the timing of traffic signals or making a specific road one-way for traffic. Through such iterations, confidence in the validity of the model can be enhanced. However, such testing becomes more difficult as models are scaled up to cover wider areas (cities, regions, nations), to cover other modes of travel, or are projected forward in time to include future economic benefits and carbon emissions. Accordingly – and remarkably – little effort is expended on demonstrating model validity compared with the enormous effort that goes into building and calibrating models. While it is generally accepted that four-stage models can be reasonably effective for projecting changes in road traffic over limited periods of time in response to investment, this may be an illusion – as the following case study of widening the M25 motorway illustrates.

WIDENING THE M25

A significant portion of the large expenditure programme devoted to the Strategic Road Network in England is earmarked for creating what are termed 'smart motorways'. In effect, a smart motorway involves creating an additional running lane by using the hard shoulder (originally intended for emergency stopping but less necessary as motor vehicles have become more reliable). Typically, a three-lane dual carriageway is increased to four lanes in each direction, as a response to high traffic volumes and the resulting congestion. This widening is accompanied by the installation of signs to set variable speed limits, to smooth traffic flows for improved journey time reliability at high traffic volumes, and to manage the consequences of crashes. Anxiety has been expressed, though, about the loss of the hard shoulder in the event of vehicle breakdown and the inadequacy of their replacement by occasional refuge areas (as discussed in chapter 1).

For one smart motorway investment, arrangements were put in place for detailed monitoring of traffic flows before the scheme opened and then at yearly intervals for three successive years. This investment added an additional lane in each direction to a section of the London orbital motorway, the M25, between junction 23 and junction 27 to the north of the capital. It was found that while the traffic flowed faster one year after opening, by year two this gain was lost on account of a substantial increase in traffic volumes (a finding that was confirmed in year three) well above that seen in nearby parts of the road network. The conclusion in the year-three report was that the results 'show that increases in capacity have been achieved, moving more goods, people and services, while maintaining journey times at pre-scheme levels and slightly improving reliability'.

When I read this, I realised that this outcome could not have been the basis on which the investment decision had been made since the orthodox approach to economic analysis supposes that the main benefits are time savings, not more movement.[5] Accordingly, I made a request under the Freedom of Information Act 2000 (which creates a public right of access to information held by public authorities) and thereby obtained a report of the traffic modelling, based on a regional model that covered a wide area well beyond the M25, as well as an economic model that converted time savings and other changes into monetary values of benefits and costs.

I found that the model substantially underestimated the amount of new traffic that arose from road widening (known as 'induced traffic'), which is why the modelled time savings were not found. The traffic modelling distinguishes between different classes of road user: principally between local users on short journeys, such as home to work; and business users on longer-distance trips, e.g. from south of the capital to the north, bypassing the centre. The forecast economic benefit was time savings to long-distance users on the expectation that added road capacity would speed up the traffic. There were forecast time savings to local users, but their economic value was almost entirely offset by increased fuel costs as drivers diverted to the motorway to save time instead of making a shorter direct trip on local roads. It was presumed, reasonably, that drivers are more aware of a time saving than they are of less visible higher fuel cost.

The traffic count evidence from monitoring does not allow a distinction to be made between the different classes of road user, but it seems likely that much – perhaps most – of the additional traffic observed in excess of the forecasts was local traffic of zero economic value, which effectively pre-empted the additional road capacity, thus negating the expected benefits to long-distance users. On the basis of the modelling, the benefit–cost ratio had been estimated at 2.9, which put the £130 million investment in the high value-for-money category and was therefore given the go-ahead. In the event, the benefit–cost ratio turned out to be very much lower, and the scheme therefore proved to be poor value.

Beyond the adverse impact on congestion, the additional traffic above forecast has further implications. Carbon emissions are higher, as are road traffic crashes, deaths and injuries. New road investments generally take credit for improved safety performance. However, some years ago I estimated the economic value of casualties from the extra accidents from induced traffic for a range of UK highway schemes and found that, on average, the value of these accidents exceeds the value of the accident savings claimed for the schemes.[6]

This M25 widening study illustrates how transport modelling can go seriously wrong, but is it a special case? Probably not. The Strategic Road Network is under greatest stress in or near built-up areas. Remote from these, the traffic generally flows freely, other than near locations such as large ports and airports, or in the case of seasonal traffic on routes to resort destinations. Near conurbations, however, local traffic competes with the long-distance traffic for which the roads are primarily intended. The question about value for money raised by the M25 case is therefore likely to apply to much of the proposed smart motorway investment. There are ten smart motorway schemes in the current five-year national road investment strategy, with an estimated benefit–cost ratio of 2.4.[7] This is, in my view, highly optimistic.

More generally, the very limited efforts being made to validate the transport models employed to inform investment decisions do not inspire confidence in such decisions. As previously noted, the monitoring of traffic volumes and speeds does not tell us which classes of users benefit and to what extent. We should adopt the travel diary techniques used by the National Travel Survey, recording all trips taken in the course of a week before a new road scheme opens and again at intervals thereafter. For a major road scheme it would be reasonable to focus on vehicle use, probably using a smartphone app to monitor the trips of a representative sample of road users. This would provide evidence of travel behaviour changing as a result of an investment, which would allow both better estimation and calibration of the model as well as indicating where and how the benefits to users arise. I find it surprising that this kind of monitoring – known as a 'longitudinal study' – has not been attempted for transport investment. Following a representative group of individuals over time is a well-established technique in health and social science research: much of the research on the impact of the coronavirus and its vaccines involves following people's symptoms and antibody levels over time, for example.

Beyond the shortcomings of modelled road traffic forecasts, transport modelling also has difficulty with other modes of travel. The economic case for London's Crossrail was largely based on the value of time savings to users of the new route and disregarded the property development that will arise, including that which has already occurred in anticipation of the line opening. This is despite it being scarcely credible that the cumulative value of these time savings would come close to the increase in value of the real estate along the route, which arises from the value of the economic worth of the activities the developments will accommodate.

A similar argument applies to HS2, the new rail route from London to the cities of the Midlands and the North that is now under construction. The economic case depended largely on time savings and was pretty marginal after substantial increases in the capital cost were recognised. Yet the economic case was silent on the distribution of benefits between London and the other cities, which of course is crucial to the strategic case for rebalancing the national economy to boost areas beyond London and the South East. Recognition of the real estate development consequences of HS2 would provide spatial content to the economic analysis, and this would of course require analysis beyond the transport silo, including investigating the plans that the cities on the route have to take advantage of the new connections.

Looking back over more than half a century of transport economics and modelling, it seems to me that a mistake was made at the outset that has distorted analysis ever since. The original simplifying assumption was that the origins and destinations of journeys could be regarded as fixed. This meant that improvements that allowed faster travel generated time savings, which could have a monetary value attributed to them. There were assumed to be no changes in access or in land use. Subsequently, it was recognised that reducing the time cost of travel would allow more people to travel, increasing demand for travel (hence the so-called variable demand models), and there was also recognition that reducing time costs could allow people to travel further. Nevertheless, travel time savings have always been the dominant component of the supposed economic benefits of investment. With hindsight, a better simplifying assumption would have been that average travel time remains constant, consistent with the findings of the National Travel Survey. Then investment that allowed faster travel would lead to changes in access and land use, which we would have had to learn how to model and value.

As it is, we are stuck in the wrong place because of what I would describe as complacent groupthink. Politicians, both national and local, tend to be keen on transport investment, hoping to boost economic growth and accommodate population growth. The Treasury accepts the transport economists' cost–benefit analysis as a good example of implementing the principle of its public investment bible, known as the Green Book. The Treasury is therefore willing to make substantial funds available to the DfT, which is happy to be a big spender, as is Highways England (recently rebranded National Highways, for no good reason), its wholly owned company responsible for the Strategic Road Network, and as are the civil engineering contractors who build new roads.

However, for an investment to be approved, it must be subject to cost–benefit analysis, which is generally carried out by specialist consultants who know what is required of them: a benefit–cost ratio of 2 or more, justifying a 'high' value-for-money categorisation. Given the complexity and opacity of transport models, the consultants usually find it possible to achieve this assessment by adjusting the model parameters, within the bounds of professional respectability. The monitoring of the outcome of an investment after opening, being limited to traffic flows, does not provide much of a check on the validity of the modelling.

Bias in modelling

What is happening in the modelling of road investments is an example of what is termed optimism bias: the tendency to skew analysis to support a desired outcome, whether consciously or unconsciously. A well-recognised example is the propensity to underestimate capital costs at the early stage of consideration of an investment. To correct for this, the DfT requires cost uplifts of up to 44% to be applied at the earliest stage of analysis.[8] Optimism bias may also arise in estimating future demand, particularly in situations where there is competitive bidding. This happened with some of the rail franchises in Britain, where fare revenue fell well short of over-optimistic projections. This meant that the franchise proved unviable and the franchisee withdrew, leaving the government to step in to maintain services. Another example concerns toll road concessions in Australia, where bids from private sector promoters based on optimistic forecasts of traffic and toll revenues led to disappointing returns to investors, who sought recompense through litigating against the scheme sponsors and their consultants.[9]

Optimism bias can result in investments whose outcomes are unintended and disappointing. One means to reduce such bias is better evaluation of investment outcomes, as discussed above, to improve the accuracy of the modelling of future projects. Another desirable approach is to improve the transparency of modelling, so that decision makers can be better informed about the biases and uncertainties of proposed investments. Transport modelling is unusually complex and opaque, with most of it using bespoke proprietary software: the opposite of open access. Other areas of public sector analysis are far more open. The Treasury's model of the UK economy is used by independent forecasters, for example. The Department for Business, Energy & Industrial Strategy has issued an energy White Paper in which it commits to accessing the best modelling expertise available in close collaboration with academia, and it has published an online carbon calculator that allows users to see how we might reduce greenhouse gas emissions.[10] The modelling of the coronavirus pandemic has been carried out not in-house by government but by a number of academic groups whose models and results are published for all to debate. And the modelling of climate change is carried out openly, collaboratively and internationally before being used in the reports of the Intergovernmental Panel on Climate Change.

The lack of openness of transport modelling does not contribute to confidence in its outputs. The transport modelling community may be characterised as an inward-looking group of experts whose work is barely subject to external critique. Indeed, a cynic might regard transport modelling as a bit of a racket. The complexity of the models precludes a deep understanding by anyone other than those professionally involved, who earn their living as modellers and who therefore have little incentive for critical evaluation of their outputs. Unlike most other professions, where professional bodies set standards and enforce adherence (as is the case for architects, actuaries, accountants, lawyers and engineers, for example), no standards for transport modellers are laid down by the profession. It has to be said that not all professions set themselves standards of performance, one prominent example being economists. Transport modellers therefore need to reflect on whether they wish to be regarded more like engineers or more like economists. As it is, they are constrained by the DfT's exceptionally prescriptive guidance for modelling prospective public sector investments – guidance on which they may be consulted but which is not their own.[11]

The engineering approach is most applicable to small-scale network models used to help make short-term decisions, e.g. to adjust traffic signals to accommodate changing traffic flows. Here, adequate data are available and the uncertainties are less significant. But as models are scaled up to cover larger areas, more modes of travel and the long timescales required for cost–benefit analysis, sufficient data are then harder to come by, uncertainty increases and the scope for optimism bias grows. The context of the modelling effort itself may induce bias, e.g. when investment decisions have already been made but which the public might challenge, or where there is competition between projects for limited funding.

Altogether, use of models to support investment decisions is both problematic and yet seemingly unavoidable.[12] One welcome recent development is that the DfT is rebuilding its two-decades-old National Transport Model to be transparent to those outside government.[13]

Having scrutinised the analytical basis for transport investment, I now consider the main current issues in investment by mode of travel. What is of central concern is the impact of those investment in the real world – in broad terms, how they help

stimulate economic growth, accommodate changing demographics and mitigate environmental harms.

Future rail investment

Understanding the impact of investment is most straightfor-
ward on the railway. Adding train capacity to the network helps
to reduce crowding. Faster trains can enlarge the travel-to-
work areas of cities, which can have economic benefits through
agglomeration, notably by creating larger and more efficient
employment markets. Faster travel between cities improves
connectivity, which can also have economic benefits, but it is
generally hard to say which of a pair of better-connected cities
is likely to benefit most. Yet rail investment is costly, and costs
are prone to overrun, which places a question mark against the
ambitions of regional entities such as Northern Powerhouse
Rail or Midlands Connect to gain government funding for new
rail capacity. What needs to be asked is whether rail investment
offers the most cost-effective approach to boosting regional
economies. Such a question would naturally be asked if regions
had devolved budgets covering the full range of possible investments, with the freedom to devise a portfolio best suited to
meeting local ambitions. In practice, though, cities and regions
have to bid for funds from central government departments
– transport as well as others – that are earmarked for specific
kinds of project. A local investment portfolio therefore largely
reflects the success of such competitive bidding, and may be far
from optimal for the locality as a whole.

The question of benefits to regional economies is crucial to
the economic case made for High Speed 2. This proposal has
been hugely controversial on account of its expense, its environmental impact and uncertainty over its economic benefits. HS2
is the most costly single British transport investment ever. Initially, the proposed initiative commanded support from across
the political parties (at an estimaetd cost of £37 billion). As plans

developed and a better understanding of the technology was obtained, estimated costs rose markedly, reaching more than £109 billion by 2020 – and the eventual cost will doubtless be even more. At the same time, the economic case was refined, yet insufficient benefits could be identified to prevent the benefit–cost ratio deteriorating to 1.5 or lower at the point when the decision to commence construction was made, the implication being that many other transport investments would have been more economically attractive.[14]

However, the problem with the economic case for HS2 is that it is based on the orthodox approach in which the main benefit is time saving to travellers. Not only is this misleading, as I mentioned in the previous chapter, but it is also silent on the geographical distribution of the benefits – a crucial consideration in a scheme intended to help rebalance the economy in favour of parts of the country beyond London and the South East of England.

I have always been agnostic about HS2 because of the inadequacy of the economic case. I do not deny that it might turn out to be a success overall. Experience of high-speed rail routes in Europe is mixed, so it is a mistake to draw general conclusions, and indeed lessons might be learned from both the successes and the disappointments. But HS2's success will depend on the efforts made within the newly better-connected cities, both to renovate and extend local transport connections to the new stations and to foster urban renewal in their vicinity. The economic case for HS2 is silent on these benefits of urban development arising from new stations and faster connections, which is another inadequacy of the standard approach to investment appraisal.

The government published its Integrated Rail Plan for the North and Midlands (IRP) in November 2021. This cut the intended section of HS2 between the East Midlands and Leeds in favour of investment in east–west links.[15] Despite headline investment worth £96 billion, public reception was mostly unfavourable.

Expectations had been excessively raised. Cities that failed to gain hoped-for improved services and new stations spoke up more loudly than the winners of this apparent lottery. Huw Merriman MP, the chair of the House of Commons Transport Committee, put it well: 'The danger [is] in selling perpetual sunlight and then leaving it for others to explain the arrival of moonlight.'

The absence of any supporting economic analysis to justify the investment choices of the IRP was noteworthy. The reason given was that 'rail schemes in the North are at increased risk of being considered poor value for money when applying conventional cost–benefit analysis. This is driven in part by smaller city populations in the North, different travel patterns, as well as the general high cost of building rail infrastructure.' The DfT's conventional investment appraisal of these schemes does not therefore provide the answer sought – which is likely why, as I discussed earlier in this chapter, the DfT has belatedly recognised that the benefits of investment take the form of better access to people and places, not the travel time savings that are the basis of conventional cost–benefit analysis.

Most rail investment is on routes between cities, yet there is a better case for rail investment *within* cities, to provide a fast and reliable alternative to cars and buses on congested roads. London's historic Underground is essential to the city's functioning, as are its surface rail commuter services. Recent new services include the Docklands Light Railway – important for opening up London's disused former port facilities for development at Canary Wharf and beyond – and the Overground, a renovated network of previously underused and neglected inner urban rail routes. A major addition to capacity in the form of Crossrail, which will be known as the Elizabeth Line, is due to open in 2022 after some not untypical delays and cost overruns. The National Infrastructure Commission, with a remit to take a long view of the nation's need for infrastructure investment, sees urban rail as a priority once the present programmes of interurban road and rail investment are completed.[16] The Commission also emphasised investment in regional rail routes rather than long-distance routes when, at the request of the government, it reviewed priorities for new rail infrastructure in the Midlands and the North of England.[17] This advice was accepted by the government in its IRP.

One aspect of rail investment that is, in principle, very desirable is electrification to reduce and eliminate dependence on diesel-powered trains and their contribution to climate change. While most of the UK rail network is electrified, the remaining parts carry relatively light traffic, much of which is freight, so the economic case for further electrification may not be attractive. Alternatives under investigation include battery power and hydrogen fuel cell propulsion.

Altogether, it is evident that the orthodox approach to the economic appraisal of rail investments, based on time savings to users, is not fit for the purpose of informing decisions makers of the real benefits of such investments.

Future road investment

The question of whether investing in transport is the best way of meeting the economic and social objectives of a region also arises with road investment. Cities and regions seek funding from central government, essentially as 'free money' to be spent on investments that can be justified as being of 'high value' in terms of the benefit–cost ratio, as discussed earlier. Yet new road capacity fills up with more traffic, consistent with the maxim that we cannot build our way out of congestion, which we know from experience to be generally true. Moreover, this extra traffic is responsible for additional detriments related to vehicle-miles travelled – particularly carbon emissions and air pollutants. Were the additional traffic to include new trips to more distant destinations, there would be economic benefit in that. It is likely, however, that much of this extra traffic comprises trips between unchanged origins and destinations, such as from home to work, with people switching to the new road capacity to save a few minutes of time at the cost of additional fuel for the longer trip. This kind of change has little economic value.

The narrative in support of road building is that it will reduce congestion and improve connectivity between cities, thereby boosting economic activity. The reality is that relief of congestion is temporary, as is the improved connectivity. Nevertheless, there is a widespread belief that more investment in road infrastructure is a 'good thing'. Both national and local politicians are keen. Every project in the road investment programme is expected to yield acceptable value, and preferably 'high value for money'. Such project appraisal is based on transport models that aim to project the future consequences of adding capacity. But as I illustrated with the M25 case study, models can be substantially wrong in forecasting the future. For this reason, we should not base major programmes of public expenditure on projections in which we cannot have sufficient confidence.

A particular problem with the current road investment programme is that it will increase carbon emissions from road traffic. What is more, the underestimation of traffic exemplified by the M25 case study we saw earlier suggests that the conflict between road investments and the Net Zero climate change objective is greater than previously supposed. In its decarbonisation plan (which I will discuss later), the DfT has undertaken to review its National Policy Statement of strategic planning for road and rail networks, which underpins its road investment programme, and to update the forecasts on which it is based to reflect developments since it was promulgated in 2014. This may point towards decreased future investment in additional road capacity, and possibly to more expenditure being directed towards enhancing resilience against extreme weather conditions associated with global warming.

In the long run, decarbonisation of road transport, and of the materials required (such as cement and steel), should deal with its contribution to global warming. In the near-to-medium term, however, conventional investment in roads runs counter to national climate change objectives. Conventional investment is largely devoted to civil engineering works: shifting earth, pouring concrete, rolling tarmac. This is labour intensive, costly and adds substantially to embedded carbon. We need to pay more attention to applying scalable digital technologies to the road network to improve operational efficiency, as I will discuss later in this chapter.

Britain has an extensive and mature road network. Adding further capacity has little overall impact. Promises of local congestion relief are short-lived, as people take advantage of faster travel to divert to the improved route, restoring congestion to its previous level. Another increasingly common rationale for new road capacity is to make land accessible for new housing on greenfield sites. But such housing generally requires the car for access – something that does not sit easily with the policy of promoting public transport and active travel towards the objective of Net Zero carbon emissions.[18]

Devolution within the United Kingdom has opened up alternative approaches to traditional road investment. The M4 motorway near Newport in South East Wales is heavily used. There had been a proposal to construct a relief road but this was abandoned on account of the cost and the likely scale of induced traffic. The Welsh government established an independent commission to consider the options.[19] The commission recommended a range of sustainable transport alternatives, which have been accepted by the Welsh government. In June 2021 the Welsh government announced it was freezing new road-building projects in advance of a review of proposed schemes by an external panel.

Overall, the UK government's current road investment programme is looking decidedly problematic in respect of both value for money and its contribution to global warming. Nevertheless, the general belief in its supposed virtues – relieving congestion and boosting economic growth – seems to be keeping this policy tanker on its traditional course. To respond to the realities, decision makers need to be better informed about the outcomes of investments in terms of how the travel behaviour of road users changes, and the resulting benefits and detriments. They also need to be better informed about digital alternatives.

Investment in air travel

In most countries, airports are owned by national or local government. Britain is unusual in having only a few airports with substantial public sector ownership (Manchester, Stansted, Luton and East Midlands Airports). Decisions about new investments are, therefore, essentially commercial. However, airport expansion requires planning consent, which allows opponents to voice their concerns. For a major airport like Heathrow, the question of whether to build a third runway has involved national debate, the outcome of which still seems some way from being settled. In response to the airport owner's wish to expand its business by constructing a third runway, the government set up an Airport Commission to consider the economic and environmental issues. The case for expansion focused on the expected growth in demand for business travel from UK exporters seeking more direct routes to distant markets, from inward investors to Britain, and to foster London as a world city in which to do business. The case against was largely environmental: both in terms of the local impact of more air traffic arising from a major international airport that is unusually near the city centre and in terms of the contribution of a growth in air travel to climate change.

The economic analysis carried out by the Airport Commission was decidedly problematic, focusing mainly on the benefits to passengers of the expected lower fares, which were not really relevant to the strategic case for expansion. Attempts to value the impact of a third runway on the economy as a whole depended on an opaque proprietary model built by a consultant, with likely optimism bias towards generating a pleasing projection of benefit.[20] The commission selected one out of three possible locations as the best bet for the third runway, but the decision was very contentious. The government was ambivalent but gave its assent to the project.

Legal action by objectors subsequently challenged the government's decision on the grounds that the implication of the 2015 Paris Agreement on Climate Change had not been taken into account. This action succeeded in the first court but was overturned on appeal. Construction of the third runway could therefore go ahead, but the coronavirus pandemic then largely wiped out air travel for well over a year, detracting from the financial viability of Heathrow's operator as well as of the airlines that use the airport. Whether air travel will revive to an extent that justifies more UK runway capacity will be considered in this book's next chapter. In any event, though, it would be a commercial decision for Heathrow's owner (a consortium of overseas infrastructure specialists and investors) whether to invest in a third runway, taking into account both future demand and the charges for airport use that might be payable by the airlines operating from the airport (the largest of which, British Airways, has expressed reluctance to pay higher charges). The return expected from the investment would have to be attractive compared with returns from other commercial investments, whether in the transport sector or beyond. In the meantime, Gatwick, London's second airport, which has a single main runway, is hoping to convert its emergency runway to regular use, which could reduce the attractions of a third runway at Heathrow.

Beyond the economic case for investment at particular airports, there is a need to consider the benefits at system level. If a third runway is not built at Heathrow, growth in demand for air travel would displace passengers to other airports. Most air travel is for leisure purposes. Even at Heathrow, only a quarter of passengers are on business trips, and if this proportion grew, it would displace leisure travellers to other London airports with spare capacity (Stansted and Luton in particular). This displacement would happen naturally in the market since business travellers would be willing to pay a premium for the convenience of a direct flight from Heathrow, whereas cost-conscious leisure travellers with more time to spare would often be willing to take a cheaper flight from a smaller airport, even if they needed to change planes on a long-haul trip.

If demand were to continue to grow, at some point all airport capacity would be in use, and fares would rise to balance demand with supply. A recent survey found that half the flights by young men living in Britain were for stag parties and a third of the flights by young women were for hen parties. If airfares for city-break weekends were rather more expensive, there might be more of a preference for Brighton or Bath than for Barcelona, with little loss of enjoyment.[21] In short, the UK airport system has the capacity to cope with growth in demand for business travel – the rationale for a third Heathrow runway – without the need to build that runway. And by not building that runway, we would avoid the cost, noise and carbon emissions that would arise.

Transport investment priorities

For public sector investment, the intention is to achieve a consistent basis for investment appraisal across the whole public sector, as set down in the Treasury's Green Book guidance for appraisal and evaluation. In practice, however, the allocation of funds is substantially influenced by history and politics. It is therefore difficult to conclude that the investment planned for road and rail will achieve a better return than investment in, for instance, fast broadband connections or vocational training. Equally, it is difficult to compare such a broad range of public investments for their contribution to strategic objectives such as boosting economic growth, accommodating population growth or mitigating environmental harms.

In particular, no attempt has yet been made to compare new digital technologies with traditional civil engineering technologies in terms of the cost-effectiveness of their contributions to enhancing access, whether that be virtual or physical. Civil engineering works are labour intensive, site specific and costly. Accordingly, we need to be aware of alternative solutions to transport shortcomings that employ digital technologies that are scalable, replicable and potentially much more cost-effective. A good example is digital signalling technologies on the railway, which allow safe operation with shorter headways (the distance between trains) and so effectively increase the passenger capacity of rail infrastructure. I will look at more examples later.

Transport investment in the public sector is mostly concerned with large expenditures on civil engineering works, yet there is much smaller-scale investment that can be cost-effective in achieving objectives: particularly in encouraging active travel through cycle lanes, low traffic neighbourhoods and pedestrian zones in town centres. The standard cost–benefit methodology allows the health benefits of cycling to be taken into account, and benefit–cost ratios for such schemes can be much higher than for conventional road 'improvements'.[22] Yet it is hardly worth the effort to undertake formal economic appraisal for small-scale investments where the main overall impact is to improve the quality of living in urban space, for which many of the benefits are difficult to quantify. Decisions on what is worth doing should be made by those who are locally elected, after proper consultation with those who are most affected.

Behaviour change

As well as investment, there is scope to improve the transport system through fostering changes in travel behaviour. Many opportunities are identified in a recent book in this series, *Transport for Humans* by Pete Dyson and Rory Sutherland, which focuses on adapting the transport system to better meet our wants and needs. In the discussion that follows I am concerned with the scope for achieving behaviour change on a scale that would have a substantial impact on the detrimental aspects of travel.

Reducing casualties

Casualties on the roads (deaths and injuries) have been a source of concern since the earliest days of motoring. In 1926, the first year for which data was collected, 4,886 people were killed on Britain's roads. The highest recorded annual death toll was 9,169 in 1941, at a time when there were only 2.5 million licensed motor vehicles in the country. Despite the growth in vehicle ownership thereafter, the number of those killed on the roads fell steadily, reaching 1,754 in 2012, when there were 34 million licensed vehicles. This downwards trend then ceased, with no subsequent improvement, although vehicle numbers have increased by a further 4 million. To those killed must be added 26,000 who were seriously injured and 125,000 who were reported to be 'slightly injured'. Car occupants accounted for 42% of deaths, pedestrians 27%, motorcyclists 19% and pedal cyclists 6%.[23]

The impressive reduction in deaths and injuries on the roads was in part due to improvements in technology – better-constructed cars, plus seatbelts, airbags and anti-lock braking systems – but much of the reduction came from better driving behaviour encouraged by enforcement of traffic laws, including legal speed limits and bans on drink driving and mobile phone use. Britain performs well in comparison with other European countries in respect of road deaths per million population. Nevertheless, the scale of deaths and injuries on the roads is too high. Enlightened authorities recognise the need for further interventions, with one example being a commitment from some to 'Vision Zero', a multinational project that aims to achieve a road system with no fatalities or serious injuries involving traffic. London has committed to this aim, with a plan for the widespread introduction of lower speed limits, low traffic neighbourhoods, safer street design, safer buses and heavy goods vehicles, plus education programmes and safety training for cyclists and motorcyclists.[24]

A specific measure that would have benefits in terms of both safety and carbon emission would be to enforce legal speed limits, which are exceeded by more than half of all cars and vans travelling on both motorways and 30 miles per hour roads.[25] A survey of motorists found that half of all respondents supported the use of average speed cameras on high-speed and medium-speed roads for this purpose.[26]

The experience of road traffic casualty reduction exemplifies the scope for changing travel behaviour to secure wider societal aims. It required political leadership, expert technical assessment of the feasibility and outcome of a wide range of measures, plus public support for the objectives and acquiescence in the interventions.

Changing prices

Another approach to changing travel behaviour is through adjusting prices. The business model of the budget airlines involves maximising seat occupancy and fare income by fine-tuning prices flexibly to match demand with supply, since an empty seat yields no revenue. Rail fares also reflect demand, although less flexibly. Ride-hailing businesses such as Uber flex fares upwards at times of high demand, both to attract more drivers and to encourage users to defer their trips until after the peak.

In contrast, the costs of most daily travel change quite slowly, but steadily. Over the ten years to 2019, the cost of motoring rose by 34% (slightly less than the increase in the Retail Prices Index), while rail fares rose by 47% and bus fares by 61%.[27] Motoring therefore became significantly cheaper relative to public transport. Much of the cost of petrol and diesel fuel for road vehicles is taxation. In 1993 the Conservative government introduced a policy to raise fuel duty faster than the rate of inflation, in part to lessen the growth in demand for road travel. The policy was initially continued by the subsequent Labour government but was effectively ended in 2000 following extensive protests from road hauliers, who were particularly negatively affected. Similarly, the *'gilets jaunes'* protests in France in 2018 were initially motivated by rising fuel prices, which were in part the result of a proposed increase in tax on the carbon content.

We can see, therefore, that the experience of raising fuel taxes to manage demand has been fraught. Motoring costs are a significant part of expenditure for many households: for car-owning households in the lowest 10% by income, average motoring expenditure is more than £50 a week[28]. For the majority of commuters who travel by car, this expenditure is unavoidable. In the long run higher fuel costs can be mitigated accommodated by acquisition of more fuel-efficient vehicles, but in the short run the impact is felt directly. While changes in fuel prices caused by the rises and falls of the oil market are generally accepted as unavoidable, tax increases may be widely resented as inequitable, meaning that this particular price mechanism to change travel behaviour tends to be politically difficult to employ.

Another mechanism to manage travel demand is to charge for using the road, known variously as road user charging, road pricing or congestion charging. We are, of course, familiar with paying to occupy road space when we park our car at a designated kerbside parking space. Toll roads, where charges are made to recover construction costs, are familiar in many countries. In general, such roads offer a faster route than the alternative on historic roads, such that the user has a choice of whether or not to pay. In contrast, road user charging involves a charge for using all roads within a defined area. An example is London's Congestion Charge zone, the main motivation of which is to reduce road traffic congestion and where all vehicles (barring certain specified exceptions) are required to pay the charge.

There has long been interest in road user charging among transport economists, who see this as a theoretically attractive means of rationing limited road space by price, justified by arguments about economic efficiency. There has recently been renewed interest in this charging mechanism, prompted by the prospect of growing use of electric vehicles, which do not incur fuel duty. Charging for road use is seen as a means both of raising revenue to fund road maintenance and construction and of managing congestion. However, the impact on congestion may be limited, based on current experience.

Only three major cities have adopted congestion charging on a significant scale: London, Stockholm and Singapore. The experience in London was a reduction of a third in the number of vehicles entering the charging zone when it was first introduced, and a substantial reduction in delays due to congestion. However, delays increased over the next five years until they again reached the pre-charge level. One specific contributory factor was a decision to make more road capacity available for bus lanes and to give space over to cycling and walking. A more general reason for reversion lies in the nature of road traffic congestion, which arises in areas of high population density and high car ownership such that there is too little road capacity for all the car journeys that might be made. Some potential road users are deterred by the prospect of unacceptable delays, so they make other choices: to take a different route or travel at a different time or by a different mode of travel where there are options, or to go to a different destination (shopping in a less congested area, for instance) or not to travel at all (such as shopping online). In this way, congestion is largely self-regulating. But conversely, measures to reduce congestion, such as congestion charging, initially reduce delays, which then attracts back road users for whom the charge is affordable but who were previously deterred by the expectation of excessive delays.[29]

It would be possible to reduce congestion to below this self-regulating level if sufficiently high charges were employed

– higher than in London and Stockholm, whose experience is similar to London's. Singapore is a contrasting case, where electronic road pricing is used to maintain the flow of traffic, with moderate charges adjusted from time to time to keep speeds within defined ranges. However, because Singapore is a city-state with no rural hinterland, it has long been policy to limit car ownership to the capacity of the road network by making only a limited number of permits available. These are auctioned off from time to time, and could recently cost up to £20,000 for a ten-year entitlement of ownership. In effect, the Singapore road charging system involves a high fixed charge to have a vehicle on the road network and a small variable charge to reflect the level of congestion. The implication of all this is that the current level of the daily charge in London is too low to have much impact on congestion, given the number of those who could afford a higher charge, including business users. However, congestion charging might be more effective in other British cities. The M6 Toll Road, which runs for 27 miles in the West Midlands, was built with private finance and charges a toll to cover costs and give a return to investors. The average daily traffic on this toll road is about half that on the nearby free-to-use M6 because local users are deterred by the charges.

More generally, road user charging can prove acceptable to the public where decent alternatives are available, whether that is an uncharged road as an alternative to a toll road or where there is an extensive public transport network, as in London. Otherwise, introducing road user charging might be problematic because of a perceived real increase in the cost of motoring, as well as concerns about equity. Road user charging is intended to increase the cost in order to reduce demand. Those displaced will be those least able, or least willing, to pay the charges. A road system that has previously been a relatively egalitarian domain therefore becomes less equal. This aspect of inequality will be reinforced by the advent of electric vehicles (EVs). The better-off purchasers of new EVs will benefit from the absence of fuel duty, while the less well-off motorists who buy their cars second hand will rely on fuel-using cars for much longer. Raising fuel duty to encourage the uptake of EVs would affect low-income motorists hardest.

In short, while the price mechanism is a logical means of changing travel behaviour to achieve wider societal objectives, it has to be deployed very carefully if public acceptability is to be preserved. An example of success is London's Ultra Low Emission Zone, which is intended to reduce harmful air pollutants from traffic by charging older polluting vehicles to enter. This initially applied on entry to the central Congestion Charge zone, but since October 2021 it has affected all vehicles within the area bounded by the North Circular and South Circular roads. Sufficient notice was given to allow London road users to replace their older vehicles, though, and there was no evidence in the mayoral election of May 2021 of voter discontent. The implementation went smoothly. Birmingham has also introduced a clean air zone in which older polluting vehicles are charged for entry, and other cities are making similar plans.

Another successful approach to charging is exemplified by the Workplace Parking Levy operated by the city of Nottingham, which aims to reduce traffic congestion by increasing the cost of car commuting. The levy is a charge on employers (currently a little in excess of £400 per space) who provide more than ten workplace parking spaces for employees. The proceeds of this initiative are invested in local transport – in particular the further development of the tram system.

In this chapter I have outlined the potential of investment to improve the transport system, which is mature and needs to be well maintained. There are certainly opportunities for improvement, mainly by bringing standards of service for public transport generally up to those achieved in the best situations – 'levelling up', as it might be termed. Yet for the most part, our travel experience on the road network would not be transformed by investment in new capacity of the traditional kind. This raises the question of whether the recent scale of road investment will be sustainable in the future, in terms of both value for money and environmental impact. A number of new transport technologies for the road network are in development or are already being deployed, however, and they are the topic of the next chapter.

This chapter has also reviewed the opportunities to change travel behaviour to reduce the harmful effects of road use. The historic reduction in road traffic casualties shows that it is possible to change behaviour to secure wider societal aims. Travel behaviour can also be changed by varying prices, but that is an approach that risks being unpopular. I will return to this later in the context of the available options for decarbonising the transport system.

Chapter 4

Do new technologies hold the answer?

The historic innovations in transport technology – railways, bicycles, cars on modern roads, motorised two-wheelers, the airliner – transformed people's lives, and there are four new technologies that are currently making – or are expected to make – an impact on how we travel: electric vehicles, digital platforms, digital navigation and automated vehicles.[1] These new technologies largely affect travel on roads. I will show that they can improve the quality of our journeys as well as reducing environmental harms, but they will not have much impact on how fast we travel. They will not, therefore, transform our lives.

Electric vehicles

Electric road vehicles (EVs) essentially involve a change of the mode of propulsion from the internal combustion engine to an electric motor plus a battery. The motivation for this innovation is to reduce transport sector carbon emissions. The leading innovator has been Tesla CEO Elon Musk, whose company developed the technology from scratch, unconstrained by a legacy of conventional vehicles. Other EV start-ups, many of them Chinese, are entering the market. New entrants face the challenge of finding a niche market in which they can create an attractive model range, establishing a brand, producing vehicles at scale and forming a sales and distribution network. They will be in competition with all the established motor vehicle manufacturers who are developing and marketing electric cars, impelled by regulatory requirements to reduce tailpipe carbon emissions and by government requirements to cease selling internal combustion engine vehicles by specified dates. The UK government has acted to end the sale of new petrol and diesel cars and vans by 2030 and hybrids by 2035. Other European governments have adopted similar approaches. These interventions in a major market are impressive, as is the general acquiescence of the manufacturers to a major policy-driven shift in technology.

The policy imperative is forcing the roll-out of new technology that is at present more costly to build than conventional vehicles. While electric propulsion is simpler and cheaper than the complex internal combustion engine, the high cost of the batteries is the reason for the higher overall cost of EVs at present. However, most new cars are purchased using leasing finance, so the higher monthly payments for an EV are offset by the lower cost of electric charging compared with highly taxed road fuel.

Better batteries

It is expected that the increasing scale of battery production – manufactured in so-called gigafactories, as pioneered by Tesla – will bring down up-front EV costs. The other main approach to cost reduction is to minimise use of the batteries' more expensive components (e.g. cobalt). Also under development are improved battery designs that reduce manufacturing costs. In the meantime, many governments offer financial support to purchasers of EVs to encourage uptake. In Britain, for example, a 'plug-in grant', with a maximum payment of £1,500, is currently available.

Demand for the minerals and metals needed for the batteries is growing rapidly as manufacturers scale up EV production. While there are anxieties about availability, it is generally the case that the magnitude of known reserves depends on the effort devoted to prospecting, which is driven by expectations of future demand. The prospects for growth in EV sales may therefore be expected to bring forward more supply of critical minerals, although the costs of new mines and the time it takes to gear up production mean that prices can be both cyclic and volatile. On the other hand, shortages of a particular material prompt innovation to use more readily available alternatives.[2] An additional consideration is that there is growing concern that we should be able to recycle battery components, which are not consumed in use.[3] On the whole, it would be premature to concluded that sales of EVs are likely to be constrained by the availability of materials in the long run.[4]

Battery performance currently limits vehicle range – the distance it is possible to travel before the battery needs to be recharged – which creates an incentive to ensure that every other feature of the vehicle minimises power consumption. Battery technology has steadily improved in respect of their key characteristics: capacity to hold charge, density of charge (energy density), rate of discharge, time to recharge, weight, cost, safety and useful life. There are trade-offs between these features for a given battery chemistry, but there is also hope that new chemistries might offer better performance than the current lithium-ion batteries.* The theoretical limit to battery energy density is unclear, unlike the energy extractable by an internal combustion engine from an oil-based fuel, the maximum of which is determined by the laws of thermodynamics. Battery technology advances incrementally, with much secrecy to protect hard-won knowledge. The eventual outcome remains

---

*For the most part, these are essentially the same as those used in small devices, with several thousand arranged in modules and packs to power a car.

to be seen. One ambition is to replace the flammable liquid electrolyte by perfecting solid-state batteries, which should be safer, lighter and faster to charge. A well-informed prediction is that improvements in battery technology will allow electric vehicles to achieve price parity with internal combustion engine vehicles by 2025–27.[5]

Core competencies

The design and manufacture of internal combustion engines has been a core competency of car companies from the very beginning. With the technology's impending obsolescence, though, the manufacturers have had to make strategic choices: whether to develop battery technology in-house or acquire the necessary expertise; whether to partner with a specialist battery manufacturer; or whether to buy in batteries, as has long been the practice for many other components, such as electricals. Any significant advance in battery technology could give a competitive advantage to the car manufacturer that has access to it, and this makes a company's strategy for battery development rather crucial.

A similar issue arises for key components such as the electric motor: should companies design and manufacture them themselves or buy them in from specialists? More generally, the question is whether the car of the future will be essentially a mechanical engineering device with ancillary electrics and electronics, as has been the case in the past, or will it effectively be an electric/ electronic device on wheels? Whichever viewpoint is adopted, though, vehicle safety standards will need to be maintained – a computer can crash and be rebooted safely; a car cannot. The traditional major auto manufacturers understand safety, but their strategic decisions are complicated by the arrival of many new entrants to the EV market (around two dozen public companies[6]) and the possibilities for vehicle automation (which I will discuss later).

Charging batteries

Potential purchasers of EVs may be deterred by concerns about the availability of charging facilities. For those with off-street parking at home, overnight charging from the domestic electricity supply allows daily travel at low operating costs. It is also likely that there will be the opportunity to export electricity from EV batteries to the grid to meet early-evening demand, recharging in the second half of the night, thereby reducing operating costs further. Regrettably, but necessarily, more front gardens will be converted into driveways. On-street charging from adapted street lampposts or dedicated charging points is increasingly common as an option for those without home facilities, but generally the costs are higher. There may be a need for regulation of electricity prices for EV charging for reasons of equity and to ensure general uptake.

Cars are driven for only about 4% of the time on average, and they are parked at home for 73%. For most, then, plenty of time is available for routine recharging. Off-street parking accounts for 75% of home-based parking, with 25% of cars parking on the street. Overall, 65% of households in Britain could provide off-street charging, although this varies by area; in London, the proportion is 44%.[7]

For longer journeys there is a need for sufficient public charging facilities to give confidence that all journeys are possible. This involves both an adequate number of rapid charging facilities and good geographical distribution of them. Provision of charging facilities on a commercial basis should become increasingly attractive as the number of EVs increases, although in the near term, government support will be needed to give assurance to prospective vehicle purchasers. In 2020 the UK government announced expenditure of £1.3 billion to accelerate the rollout of charge points for electric vehicles in homes, on streets and on motorways across England. Yet substantially more investment may be needed if the phasing out of conventional cars is to be achieved on the intended timescale. One estimate is that the UK will need 400,000 public charging points by 2030 – a very significant increase over the 35,000 that existed in 2021.[8] Transport for London (TfL) has projected that, with sufficient charging infrastructure, 46% of the distance travelled by cars in London could be by electric cars by 2030.[9]

One seemingly attractive solution to the problem of 'range anxiety' on longer trips is the 'plug-in hybrid' (PHEV): a vehicle with both an electric motor and an internal combustion engine. With these vehicles, any lack of en route charging facilities does not thwart the journey but home overnight charging allows low-cost daily travel. However, PHEVs are problematic. Because they have both a conventional engine plus fuel tank and electric propulsion including a battery, they are heavy and offer only short-range electric propulsion. In practice, they tend to be used substantially in non-electric mode, so that carbon emissions in real-world driving are some twofold to fourfold higher than in the lab tests used to approve vehicle performance.[10] Given this performance in practice, it seems that PHEVs are not a useful bridge to full battery electric mode.

It is not enough to electrify just cars

While the present intention is to phase out the sale of conventional cars and vans in favour of EVs, this policy does not carry across straightforwardly to heavier vehicles, for which there are a number of options for decarbonisation. Electric buses are now widely used on routes for which overnight charging at the depot is sufficient. For heavy goods vehicles (HGVs), a constraint on electrification may be the weight of sufficient battery capacity to allow large loads to be moved over useful distances. Nevertheless, a number of manufacturers have electric versions in development or already being marketed.

An alternative to battery-based electric propulsion is a fuel cell that uses hydrogen to generate electric power. Fuel cells require platinum as a catalyst but do not need any other materials that might be supply-limited. The constraint in this case is the availability of hydrogen. The 'greenest' hydrogen is generated by the electrolysis of water using electricity from renewable sources, although this is costly. Hydrogen from natural gas is cheaper but generates undesirable carbon dioxide as a by-product, the capture and geological storage of which would add considerably to the cost. Beyond generation issues, the distribution of hydrogen to refuelling facilities also adds to the cost of supply. Nevertheless, trials of hydrogen fuel cell buses are underway in Aberdeen, Brighton and Birmingham.

Electrification of road vehicles will not in itself decarbonise road transport: it will also be necessary to decarbonise the generation of electricity, relying on renewable and nuclear sources. The United Kingdom has almost eliminated coal burn at power stations. Eliminating gas will be more difficult, but that will need to be achieved if we are to reach the net-zero objective. The adoption of EVs will result in an increase of 20–30% in the required electricity supply. There will also need to be a corresponding strengthening of the electricity distribution network, although effectively managed overnight charging will help limit peak demand on the system.[11] Generally, achieving an adequate electricity system for EV charging seems possible, but investment will be required, and that investment will need to be recovered over time from vehicle users. While the electric charging system is ramping up, the demand for oil-based fuels will decline and, in turn, so will the number of retail outlets at which such sales remain profitable. This is a consideration that might encourage more rapid adoption of EVs.

Manufacture of EVs and their batteries currently involves processes that liberate carbon, hence the notion of 'embedded carbon'. It is argued that this should be taken into account when assessing the magnitude of carbon reduction that results from electrification of road transport. A similar argument applies to the maintenance and construction of the road network. However, the net-zero objective implies the elimination of liberated carbon from all manufacturing processes by 2050. Although this will not be easy to achieve for key materials such as steel, cement and plastics, efforts are underway to devise innovative solutions to this end.

The switch from oil-fuelled vehicles to electric ones will be transformational for the auto industry and will require substantial new investment by the electricity supply industry, for both generation and distribution. Both industries seem to be rising to the challenge of the business opportunities that decarbonisation is creating.

The shift to EVs will not be transformational for drivers, however, as the change of propulsion will not change the speed of travel. Cheaper motoring is possible because petrol and diesel for road vehicles are highly taxed whereas electricity is not. But the distance we travel is limited by the hour a day of average travel time and the speed of travel, so lower operating costs would not have much impact on overall vehicle use. Moreover, it is possible that a charge for road use will be introduced to compensate for the loss of revenue from fuel tax – something we will discuss later.

Digital platforms

Digital platforms create virtual markets. Physical markets require buyers and sellers to be in the same place, whereas virtual markets take advantage of digital technologies to allow us to deal remotely. The internet has fostered huge growth in online retail, which reached 20% of all UK sales prior to the pandemic and got as high as 36% during it. Digital platforms have been very successful in helping us to make travel arrangements, booking air and rail travel as well as hotel accommodation.

Hailing a taxi

The most striking application of digital platforms to travel has been to make it easier to find a taxi, with Uber being the best-known provider. In the past, you would have had to take your chance to flag down a licensed cab in the street, joined the queue at the station taxi rank or booked a minicab (a private hire vehicle, or PHV) by phone, with sometimes ambiguous promises being made about how long your vehicle would take to arrive. Booking online by means of a smartphone app provides visible assurance about arrival time at your pickup point, and it tells you who is driving the cab and what the fare till be in advance of the trip. It also gives you the ability to add a tip, pay automatically via a credit card and rate the performance of the driver. Fares are generally competitive with licensed cabs ('black cabs' in London), but they flex to match supply with demand at times of peak usage by adding a surcharge to the regular fare (which licensed cabs cannot do). This both attracts more drivers and encourages some users to defer their journey to a less busy time or to use an alternative mode of travel.

Generally, ride-hailing – as the service provided by Uber and other similar businesses is generically known – has improved the efficiency of taxi services and has therefore proved very popular. Prompted by the success of Uber, London Black Cabs are available via the Gett app. As well as providing door-to-door taxi services for individuals, the ride-hailing companies offer shared use at lower fares for those travelling in the same direction, although this was suspended during the pandemic. Prior to the pandemic, pooled trips accounted for up to a third of ride-hailing trips in US cities where the service was offered. However, shared use means unpredictable, and longer, journey times for some of the passengers.[12]

The growth in ride-hailing services has raised a number of significant issues, one of which is the impact on urban traffic congestion. As a readily available service, ride-hailing might add to traffic; on the other hand, though, it may replace trips in personal cars, particularly where parking availability may be a limiting factor. Evidence of outcomes is mixed and varies from city to city. One study of traffic in San Francisco found an increase of 62% in 'vehicle-hours delay' between 2010 and 2016 compared with an estimated counterfactual increase of 22% without ride-hailing.[13] The impact of ride-hailing in London, largely as Uber, is indicated by the growth of licensed PHVs from 50,000 in 2009 to 87,000 in 2019 – PHVs now account for half of all car traffic in central London. Nevertheless, car traffic in central London has been on a declining trend, which does not suggest that ride-hailing has added to congestion.[14] The conclusion of TfL is that growth in PHVs in London has largely been a matter of substitution for other vehicles and has not contributed directly to increased congestion.[15]

A further issue is the impact of ride-hailing on public transport use. It might compete by, say, attracting customers from buses, or it might complement by offering 'last mile' journeys, e.g. from suburban stations to home late at night, thereby providing an alternative to using a car for the whole journey. Evidence from a range of cities about substitution and complementarity is mixed, although ride-hailing is not generally used for commuting; rather, evening trips after the rush hour predominate.[16] Paradoxically, there has been stability or a slight decline in per person taxi/PHV use in recent years in both London and the rest of England, despite a large increase in the supply of ride-hailing and other PHVs, which does not suggest any general switch from buses to ride-hailing.[17]

More generally, while questions are properly being asked about how the growth in ride-hailing is affecting both congestion and public transport use, few cities have concluded that the impact is sufficient to justify significant mitigating measures. An exception is New York City, which in 2018 introduced a cap on the number of ride-hailing vehicles permitted to operate as well as attempting to limit the amount of time the vehicles spent cruising without passengers.

Drivers' working conditions

Beyond the impact on traffic, a further concern is the working conditions for those who drive for ride-hailing companies. Generally, the approach of the companies has been to regard drivers as independent contractors who offer their services to passengers for journeys that are booked through the app. As independent contractors, they provide their own vehicle, incur the operating costs and pay a commission to the ride-hailing business (typically 25% of the fare). Given their independent status, they have not been entitled to the benefits that would be available to employees, such as the legal minimum rate of pay, holiday pay and pension entitlement. Driving for ride-hailing firms is an 'unskilled' occupation, in that it requires only a driving licence and evidence of good character, and accordingly in a competitive market earnings tend not to be much above the minimum wage.

The legal position has been changing. In 2020 California's state legislature enacted a requirement classifying ride-hailing drivers and other gig economy workers as employees. However, this was overthrown by a referendum on a proposition that app-based transportation companies should be an exception and be allowed to classify drivers as independent contractors – something the ride-hailing companies strongly advocated. However, the constitutionality of the referendum is a matter of ongoing debate in the courts. The position in California contrasts with the situation in Britain, where a series of legal actions by Uber drivers eventually reached the Supreme Court, which ruled that the drivers were 'workers'. This employment category sits between 'employee' and 'independent contractor' and provides an entitlement to holiday pay, sick pay and the minimum wage.

A key question concerns whether drivers are 'at work', and therefore earning, only when they accept a paying passenger – which is what the ride-hailing operators would much prefer – or whether they are at work and earning any time they make themselves available for work, as the Supreme Court required. If the latter applies, operators will be reluctant to pay drivers when there is insufficient demand from passengers, with the result that drivers would not be free to clock on when they wished. This would imply some kind of zero-hours contract arrangement, where the operator would decide when work was available, rather than the previous flexibility for the driver to decide when to work. In 2021 Uber recognised the GMB, an established trade union, as being able to represent drivers in the United Kingdom. This is the first time a union has been given recognition by a ride-hailing business, and it may represent a change in their approach to their workforce.

A further question that arises concerns the economic viability of the ride-hailing model in different markets with different regulatory regimes. Uber and others are able to employ sophisticated analytics to optimise performance and financial returns, yet the twin connected markets in which they operate – for passengers and for drivers – are both very competitive, and it can be hard to recover fixed costs and, therefore, to generate sustainable profits to reward investors.[18] One factor is the high salaries paid to staff other than drivers compared with the taxi sector generally. A related issue is the liability for value added tax (VAT), which UK businesses with annual sales of less than £85,000 are not required to add to their charges – in practice, this exempts owner–drivers. But if Uber is now seen as the provider of taxi services, it may be required to charge VAT on fares, which would tilt the playing field against ride-hailing. A recent call for evidence by HM Treasury on VAT in relation to digital platforms did not point to any likely policy change.[19] However, a recent court decision stated that a licensed taxi operator such as Uber is entering into a contract with a customer rather than acting as an agent for the driver.[20] Uber now adds VAT to its fares, and similar businesses providing online or telephone bookings may need to follow suit.

Despite such challenges to the finances of ride-hailing businesses, the combination of the convenience for users and the operational efficiency of the taxi summoned by smartphone app is likely to sustain the business model, even if fares have to increase to match or exceed those of regular taxi services. The general availability of apps to summon taxis of all kinds means that this digital platform technology has become generic. Uber has lost its first-mover advantage and now has to compete with others in the market in respect of cost and service, to the benefit of customers. Yet clearly there is a need for reform of the legislation governing taxis and PHVs to recognise the new business models made possible by the digital platform technologies.

Mobility through shared use

Digital platforms are also the basis of other transport innovations that allow shared use of vehicles – such as electric bikes (e-bikes) and e-scooters – where the vehicle is located and paid for via a smartphone app. Dockless conventional bikes were initially popular but have subsequently faded from the scene. There is currently quite a lot of enthusiasm for innovative types of 'micro-mobility', as it is known, that involves electric propulsion. One aspiration is to meet the need for 'first/last mile' travel between a public transport terminus and home, which can make trains and buses more attractive alternatives to the car. There are a number of trials of e-scooters underway in Britain. One issue is their safety on roads, given that they are not suited to operating on pavements, although it is arguable whether their users are any more vulnerable than bike riders.

Digital platforms based on smartphone apps have also made possible various forms of car sharing (beyond traditional rental by the day or longer). Pay-per-trip car clubs, which may be commercial or co-ops, offer an alternative to personal car ownership. Trips may be back-to-base or one way. Sharing of car rides (often known as lift sharing) is another possibility, where people making the same journey can identify each other, for instance for journeys to work or for long-distance trips. The scope for car sharing is limited by reluctance to share one's car with others; a recent survey during the pandemic found that 80% of respondents were concerned about sharing a journey or a car with someone they did not know.[21] A number of commercial car sharing initiatives have proved unviable, in part due to competition from ride-hailing, which is in effect a chauffeured form of shared use.

Ride-hailing is a particular example of what is known generically as demand responsive transport (DRT), something that has had a long history in the form of minibuses prebooked to meet the needs of those with disabilities (branded as Dial-a-Ride in London and elsewhere). DRT minibuses based on a smartphone app are more flexible and user friendly than booking the previous day (as required by Dial-a-Ride), and they are currently being trialled in a number of locations in the United Kingdom and elsewhere. The economic viability is problematic, though, with many such DRT ventures having closed mainly on account of high operating costs.[22] Given that much public transport cannot cover its full costs and therefore receives subsidies, commercially operated DRT services may be disadvantaged. However, some local authorities are adopting DRT to replace fixed-route bus travel that they have been subsidising to fulfil social objectives.

A multimodal demand response service known as Mobility as a Service (MaaS) has been devised to provide an attractive alternative to car ownership.[23] The offering on a single platform might include buses, trains, taxis and car rental, with a range of packages for a fixed subscription (analogous to subscriptions for mobile phone usage). What is unclear is how attractive such an approach would be to users – over and above what is available in cities that have an integrated bus and rail system with payment by means of a contactless card, a ride-hailing taxi service, and a comprehensive wayfinding app that covers all modes of travel (such as Google Maps or Citymapper). Also unclear is the attractiveness for private sector operators and investors, given the complexity of the offering and the associated operational and financial risks. Public transport operators generally value their direct relationship with customers and the direct cash flow that results, so they would probably be reluctant to accept a private sector MaaS lead. A public transport operator would be better able to take on the risks as the MaaS lead but might be disinclined to do so in case they lost passengers to other modes. Nevertheless, the Jelbi multimodal app is offered by the Berlin public transport operator, based on the technology of Trafi, an MaaS startup. The financial viability of such offerings has yet to be demonstrated, however.

A number of other pilot implementations of MaaS have taken place, with the public policy aim being to offer a sustainable alternative to individual car ownership. A small but well-documented pilot operated in Sydney, Australia, between 2019 and 2021 (when the pandemic cut it short). The bundles of services offered for a fixed fee subscription were developed over the course of the exercise as user preferences were established. One finding was that 15% of participants reported that their experience changed their view of car ownership, although how this might affect actual ownership was unclear. As an alternative to city-wide MaaS deployment, there may be opportunities for localised provision aimed at reducing car use, e.g. for larger employers or for new housing developments.

Economics of shared use

The major impacts of digital platforms in the transport sector are seen in the rapid growth of ride-hailing (Uber operates in more than 600 cities around the world) and of online booking of rail and air travel. These changes have been disruptive to the businesses of established operators. In contrast, growth in car sharing, micromobility, DRT and MaaS have been relatively slow, indicating that while niche markets are being developed, disruption to established travel modes seems unlikely. These services seem difficult to establish as viable businesses that can attract substantial investment for fast growth, and it therefore seems unlikely that they will prove to be significant alternatives to conventional public transport.

Some transport analysts believe that car sharing will become an important means of reducing both traffic congestion and road vehicle carbon emissions.[24] In my view, those that have high hopes for car sharing are likely to be disappointed. One reason for my scepticism is the difficulty of establishing a viable business model.[25] Another is that personal car ownership seems likely to remain attractive. It is often argued that because cars stand idle for 95% of the time, car sharing must be economically attractive since the capital costs would be spread over much more mileage. But this is not unique: my washing machine is unused for at least 95% of the time; I could share with others at the launderette; but it is more convenient to have my own appliance. The fact that owners are willing to pay substantial sums for sole use attests to the value they place on the convenience of having their car available when they want to use it. There are also 'feel good factors' that motivate people to buy cars (to be discussed later). A study of car ownership in four US cities found that the value experienced by owners exceeded the costs of ownership and operation, and that the value from feel good factors exceeded the value from making journeys.[26]

More generally, the success of digital platforms depends on 'network effects', whereby some services become more valuable to each of their users as more people use them. The telephone is one historic example; WhatsApp a recent instance. Network effects reflect demand-side economies of scale, where benefits to users – the source of demand – grow as scale increases. In the transport sector, we are concerned with platforms that function 'online to offline' – digital access to physical mobility. As well as matching users with services, the platforms optimise operations, e.g. by selecting the fastest routes and predicting the location of future demand. The negligible cost of digital scaling means that these platforms can handle huge volumes of information – about user preferences, the availability and price of services, payments, etc. In the past such data handling would have been limited to large organisations, but now the availability of cloud computing with unlimited capacity helps innovators enter the market, scale rapidly and compete aggressively, provided that investors can be persuaded of the ultimate profitability of the offering.

Demand-side economies of scale can grow much faster than costs. However, the main challenge for transport digital platforms arises because the supply side involves physical plant and infrastructure whose economic availability is finite, meaning that capacity – a perishable commodity – must be carefully managed. An important tool is revenue management, where varying prices are used to match supply with demand. This needs lots of data and lots of supply and demand to run well. Yet it is the costs of supply that limit the profitability of Uber and other service providers.

From the users' perspective, digital platforms can improve the quality of the travel experience but they do not make much difference to the speed of travel, and they will therefore not have an impact comparable to the major advances in transport technology of the past.

Digital navigation

A combination of recent advances in a number of digital technologies is having a significant impact on travel behaviour on the road network by providing route guidance that takes account of traffic conditions. What may be termed 'digital navigation' involves the use of satellite navigation (satnav) to provide spatial positioning to high precision, digital mapping, the ability to detect vehicle speeds and hence the location of traffic congestion, and routing algorithms to optimise journeys.

Digital mapping has essentially digitised traditional paper maps, in the same way that digital documents have digitised their paper versions. However, digital maps have the capacity to accommodate vast amounts of data, including from satellite images and street level information – and what is more, that data can be readily updated. The combination of satnav location and digital mapping provides a navigation service. For shipping and aircraft, this takes the form of a recommended course between waypoints. For road vehicles, turn-by-turn route guidance is provided. And for travellers within cities, smartphone digital journey planner apps offer routes that depend on the mode of travel: car, taxi, public transport, walking or cycling.

On the roads, satnav location allows the progress of vehicles to be monitored and congested conditions to be detected. Real-time congestion is included in the most useful route guidance offerings – an approach pioneered by Waze, originally an Israeli tech start-up. This approach uses smartphones to collect crowd-sourced data about traffic conditions from users, including slow motion in congested traffic, and feeds back suggestions to users about routes with the shortest travel time, taking account of traffic conditions. The apps also provide estimated journey times in advance of setting out. Waze became a subsidiary of Google in 2013, and Google now employs the Waze technology for route guidance in its Google Maps offering. These smartphone apps are free to use, with the cost of the service being funded by retailers paying to have the map show their locations. Other providers of route guidance technology that take account of traffic conditions and estimate time of arrival include TomTom and Garmin, both of which operate via dedicated devices that are either stand-alone or are fitted to the car.

While digital navigation is in widespread use by road users, remarkably little information is publicly available about performance, and particularly about how routes are optimised, the suitability of recommended routes, the accuracy of estimated journey times, and the impact on the functioning of the road network as a whole. Nevertheless, there is evidence to indicate an impact on the use of minor and major roads, and an impact on traffic congestion and the optimisation of the road network.

Growth of traffic on minor roads

Recent revisions to British road traffic statistics appear to show that there has been a substantial growth of motor vehicle traffic on minor roads in recent years, from 108 billion vehicle miles (bvm) in 2010 to 136 bvm in 2019: an increase of 26%. Traffic on major roads rose from 197 bvm to 221 bvm over the same period: a 12% increase.[27]

Road traffic statistics are based on a combination of automatic and manual traffic counts. Traffic on all major roads is counted on typical days. However, given the vast number of minor roads, it is only possible to count traffic in a representative sample of locations every year, and the observed growth in that sample is extrapolated to estimate minor road traffic overall. Estimates from a fixed sample may drift over time if the sample becomes less representative of the minor road network as a result of changing patterns of use. To account for any errors incurred in the fixed sample, the sample is revised every decade through a benchmarking exercise involving a much larger sample of locations.

The most recent minor roads benchmarking exercise was published in 2020, based on 10,000 representative locations.[28] Overall, the benchmark adjustment for 2010–2019 was 1.19, which is the factor to be applied to the 2019 data set from the original sample locations to bring this to the observed traffic level based on the benchmark sample. Data for minor roads traffic for intermediate years are adjusted pro rata, to avoid a step change in the reported traffic data. There is significant regional variation in the adjustment factor: from 1.35 for Yorkshire to 1.09 for the East of England, with London at 1.32. Within the class of minor roads, there are B roads, for which the factor is 1.25, and

C roads, for which it is 1.17; while for urban and rural roads it is, respectively, 1.22 and 1.15. Applying the regional weightings yields an overall increase in traffic on minor roads of 26%, as noted above, whereas the increase based on the original sample would have been 6%. The previous benchmarking exercise published in 2009 found a smaller overall adjustment factor of 0.95, with a regional range of 0.81–1.08.

The substantially greater adjustment required following the recent benchmarking compared with the earlier exercise suggests that there has been a significant real change in the use of minor roads, beyond errors arising from drift in the sample. Importantly, had the increase in minor road use been even across the country, the traffic estimation based on the sample would have been close to that from the benchmark exercise. The major difference between sample and benchmark therefore suggests considerable variation of minor road traffic growth – variation that is not understood.

One factor that is contributing to the growth in traffic on minor roads is the increase in van traffic, including that arising from the growth of online shopping with home deliveries. The number of vans (light commercial vehicles) registered in Britain increased by 28% between 2010 and 2019.[29] Total van traffic increased by 34% over the same period, with an increase of 49% on urban minor roads compared with 10% on urban A roads, although 'delivery/collection of goods' was less important in respect of journey purpose than 'carrying equipment, tools or materials'.[30] Despite these increases, though, van traffic amounted to 15% of traffic on urban minor roads in 2019, and 19% on rural minor roads, with cars being responsible for 82% and 78% of traffic, respectively. The growth of van traffic on minor roads is therefore responsible for only part of the overall traffic growth on these roads.

The changed distribution of traffic on minor roads might have arisen as a result of intentional interventions aimed at reducing such traffic. It has long been the practice to discourage 'rat running' on urban minor roads by means of suitable physical control measures. The creation of street environments support-ing a shift from private car use to active travel – walking and cycling – has been an aim of recent policy initiatives to create low-traffic neighbourhoods. In response to the Covid pandemic, the government issued statutory guidance that local authori-ties in areas with high levels of public transport use should take measures to reallocate road space to walking and cycling, in order to to reduce crowding on buses and trains. Such meas-ures were intended to reduce traffic in certain locations while possibly increasing it in others through diversion. However, the net effect of intentional interventions would be to reduce traffic overall, so this cannot account for the reported growth of traffic on minor roads.

My conclusion is that the most likely main contribution to the large growth in traffic on minor roads is the widespread use of digital navigation, which makes possible the general use of minor roads that previously were largely confined to those road users with local knowledge (as well as extending such local knowledge). Diversion to minor roads is likely to occur when and where major roads are congested, and it represents an effective increase in the capacity of the road network. This means that it generates additional traffic – analogous to the increase in traffic seen when major roads are widened, as discussed earlier. Diver-sion arising from congestion on major roads may be a reason why the recent increase in minor road traffic has varied con-siderably across the country, as noted above, with the largest increases likely to occur on minor roads adjacent to congested major roads.

Diversion to major roads

As well as encouraging the use of minor roads, digital naviga-tion may divert traffic from local roads to roads intended for longer-distance traffic. One case where such diversion may have occurred is the widening of the M25 (London's orbital motorway) between junctions 23 and 27 to the north of the city, which I discussed earlier. There was substantial growth in traffic after the widening, above the level that had been forecast, much of which (most, even) arose from diversion of local trips, such as from home to work, to take advantage of faster travel on the motorway, despite the greater distance and higher fuel costs incurred.

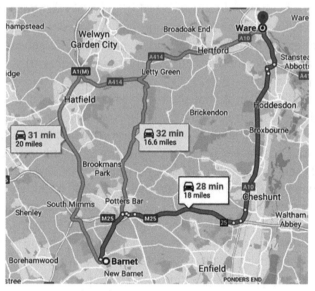

Figure 4.1. Google Maps screenshot, Barnet to Ware, weekday mid-morning.

The contribution of digital navigation to facilitating such diversion cannot be inferred from the available data, but it is certainly plausible. This can be illustrated by the example in figure 4.1, which shows a journey from Barnet in north London to Ware in Hertfordshire, a distance of about 16 miles on the most direct route via lesser roads. However, the fastest route involves diversion to the M25, saving four minutes. Regular users of digital navigation would have up-to-date information for each journey, while irregular or non-users would probably also be aware that use of the M25 would offer the fastest journey.

The M25 case study suggests that local traffic may be expected to take advantage of the capacity increase of major routes in the vicinity of urban areas that generate lots of traffic. These are the locations where the Strategic Road Network is under greatest stress and where investments to increase capacity are thought to be most needed. However, this local traffic negates the benefits expected for long-distance road users and therefore undermines the economic case for investment. The growing use of digital navigation would tend to contribute further to weakening the case for such investment.

Mitigating congestion

While the M25 case study is an illustration of the maxim that we cannot build our way out of road traffic congestion, the development of digital navigation nevertheless probably offers the best means available to mitigate the impact of congestion. Congestion arises in or near areas of high population density and high car ownership, where the capacity of the road network is insufficient to cope with all the trips that might be made. Drivers are deterred by the prospect of time delays and therefore make other decisions: to travel at a different time, by a different route, by a different mode, to a different destination (where that is an option, such as for shopping), or not to travel at all (by shopping online, for instance). Congestion is therefore substantially self-regulating, in that if traffic increases, delays worsen and more potential users are deterred on account of the time constraint.

Digital navigation that takes account of congestion in real time can offer less congested routes, thereby making better use of the existing road network and reducing road users' exposure to congestion. One problem that may arise is that traffic might be diverted onto unsuitable roads – local environments and neighbourhoods may be adversely affected, or large vehicles can even become stuck. Some satnav systems are designed specifically for heavy goods vehicles, and these avoid use of unsuitable routes, taking account of vehicle size, weight and load. However, it is unclear how many truck drivers use a bespoke system, which can be expensive, and how many rely on more popular satnav devices designed for cars or free-to-use options on a smartphone. Diversion onto unsuitable routes is a problem that could be mitigated through collaboration between digital navigation providers and road authorities (this will be discussed later).

Beyond the rerouting of traffic to less congested roads, there is a feature of digital navigation that mitigates the unwelcome experience of traffic congestion: the prediction of journey time, or estimated time of arrival (ETA). When road users are asked about their experience of congestion, both in surveys and in discussion, their responses indicate that uncertainty over journey time is a more important adverse consequence than lower speed. One survey of road users in England asked about priorities for increased expenditure on motorways: almost half of respondents ranked improved flow and reduced congestion as a priority, compared with less than a quarter wanting reduced journey times. Accordingly, an important benefit of digital navigation is the forecast of ETA in the light of prevailing traffic conditions on the selected route – something that reduces journey time uncertainty substantially. There is a lack of publicly available data on performance, but anecdotal experience suggests that ETAs can be reasonably reliable, at least for shorter journeys. It has recently been demonstrated that predicted journey times from Google Maps can be improved by the application of the graphical neural networks developed by DeepMind, the innovative AI business that is associated with Google.[31]

Who is in charge of the road network?

While diversion onto less congested routes may be helpful for users of digital navigation, there is a question over whether this is optimal for users of the road network as a whole. Digital navigation employs proprietary algorithms whose performance cannot be assessed externally. An algorithm might respond to a build up in congestion by diverting all traffic to a single alternative route until that became congested, repeating the process to spread traffic across available routes until congestion abated – or the algorithm might spread traffic across all available routes at the outset. The algorithm might also anticipate the build-up of congestion based on historic experience. But in any event, the routing algorithm used by one provider would not take direct account of the activities of another provider. The providers of digital navigation services are very secretive and there is almost no published information on their design or performance.

Having journey time predictions before setting out on trips may increase the operational efficiency of the road network. To a first approximation, there are two kinds of road user. The first is those who need to be at their destination at a particular time, whether to get to work or to a meeting or to deliver time-critical goods; these users are able to use predictive journey time information to decide when best to set out. In contrast, the second kind of users are more flexible in timing their trips, e.g. for shopping, leisure activities or delivering goods within wide time slots; this group could use estimated journey times to minimise their exposure to congested traffic. This is potentially win–win, in that the more the flexible road users can avoid peak traffic, the less traffic there will be at peak times for those who are not flexible.

When only a small number of people were using digital navigation tools that took account of congestion, there were benefits for those diverted onto less congested routes as well as for those that did not divert (because they experienced rather less traffic). But as more drivers use route guidance as it currently exists, and as diversion becomes more common, the net benefit to society is less clear. Nevertheless, there is potential for digital navigation technologies to make a significant impact on how the road network functions, with implications for users beyond those equipped with the technology, and this is a reason for collaboration between providers of navigation services and road authorities to achieve the best outcomes for all road users and for those exposed to traffic in their local environment. There is therefore a case for a regulatory regime to govern the operations of digital navigation providers.

The road system is well regulated to achieve safety and efficiency. Vehicle types must be certified as safe, and (in Britain and some other countries) older vehicles are tested annually; drivers must be licensed and are penalised for traffic offences; and the road infrastructure is built to prescribed standards. Given the potential scale of the impact of digital navigation devices on network operations, a licensing regime would arguably be appropriate for providers. This might require information to be exchanged with road authorities, guidance to be accepted to avoid adverse environmental and social impacts, and mutual collaboration to optimise the operational efficiency of the network as a whole while at the same time optimising the experience of individual road users.

Relevant legislation does in fact already exist in Britain: the Road Traffic (Driver Licensing and Information Systems) Act 1989 requires dynamic route guidance systems that take account of traffic conditions to be licensed by the government. The licence could include conditions concerning the roads that should not be included in route guidance and a requirement to provide road authorities with data on traffic conditions. The intention of the legislation was to facilitate the introduction of a pilot route guidance system that had been developed by the government's Transport and Road Research Laboratory, although in the event this was not taken forward. In practice, this legislation has been disregarded since no licences have been issued.

It is essential for city authorities that operate urban traffic management systems aimed at optimising traffic flows by varying the timing of traffic signals to have access to good traffic data. In London, where three-quarters of congestion is the result of excess demand and one-quarter results from planned events or unplanned incidents, 75% of the 6,000 sets of traffic signals continuously vary their timing to optimise flow across both individual junctions and the network as a whole. This reduces delays at junctions by about 13%. The benefits from such dynamic traffic signal technology were demonstrated during the 2012 London Olympic Games when major changes in flow were managed successfully.[32]

Voluntary data-sharing arrangements exist. In London, TfL makes data freely available to app developers and is a partner in Waze's Connected Citizens Programme, which provides cities with real-time data on traffic disruptions. Uber also provides anonymised data on traffic movement for many cities, based on data from its fleet of vehicles. However, such data sharing does not appear to extend to national road networks.

It seems likely that digital navigation could be managed to achieve better outcomes, not just for users of the devices but for all road users. This would involve collaboration between road authorities and the providers of navigation services. The aim would be to avoid the use of unsuitable roads and to optimise network performance at times of stress, whether due to daily congestion, major incidents or adverse weather conditions. This should not affect the business models of the providers, which depend on either selling map location and direction finding to retailers, or selling navigation services that function with the hardware fitted as an integral part of the vehicle to car manufacturers.

Optimisation of the functioning of the road network through more effective use of digital technologies offers an alternative to costly civil engineering to increase capacity, particularly since the costs of digital navigation are not borne by the public sector.

Indeed, the economic benefits of civil engineering investment in additional road capacity are likely overstated on account of the diversion of local traffic of little economic value to take advantage of increases in major road capacity, thus negating the economic benefit to long-distance users (as the M25 smart motorway case study discussed earlier illustrates). A prospective economic analysis of the future UK highways investment programme estimated that ten planned smart motorway investments would have an average benefit–cost ratio of 2.4,[33] yet this could be very substantially overstated on account of traffic diversion facilitated by digital navigation.

If it proves possible to use digital navigation to modify driver behaviour to achieve better outcomes for all road users, this would be an example of a 'nudge': an intervention that structures the choices available in such a way as to help individuals make better choices without restricting their freedom to choose. Richard Thaler, Nobel laureate economist and proposer of the nudge concept, cites satnav technology on smartphones as an example: you decide where you want to go, the app offers possible routes, and you are free to decline the advice if you decide to take a detour.[34] However, for nudging to succeed, we would need to be confident that there is indeed a benefit to all road users, whether nudged or not.

Automated vehicles

The new transport technology that has the potential to have the biggest impact is the automation of road vehicles (also known as autonomous vehicles (AVs) or driverless cars). Automated vehicles are familiar on new railway routes such as London's Docklands Light Railway, which was purpose built for driverless operations. In contrast, historic railways are difficult to automate, as are roads.

Why has there been so much excitement about automated vehicles? It is very largely an example of technology push, rather than consumer pull. A variety of technological advances in combination allow the human driver to be dispensed with – under specified conditions, at least. The human is replaced by a robot driver.* Many are optimistic about the prospects for driverless cars; others are sceptical, including Christian Wolmar, whose book in the series in which this book also appears is titled *Driverless Cars: On a Road to Nowhere?*[35]

There are a number of technological components that go into making the robot driver. The robot must sense its surroundings using video cameras, usually plus radar and often plus lidar, which uses the reflection of laser light to detect objects. It must know where it is located in relation to the features of the road network, requiring satnav location and high-definition three-dimensional digital maps, which need frequent updating. Fast software programming is required to fuse all images, using inexpensive hardware with minimum power requirements. Unlike a factory robot, the robot driver cannot be pre-programmed to deal with all situations that might arise, so the robot must learn on the job by utilising artificial intelligence.

Approaches to automation

There has been extensive development of AV technology both by new entrants to the road vehicle sector and by established manufacturers. There are broadly two approaches. The *evolutionary approach* adds individual features in order to reduce the task of driving: things such as cruise control, lane keeping, lane changing and automated valet parking. These can be seen as further advances in the sequence of developments that have delivered established technologies such as automatic gear shifting and automatic emergency braking. The potential problem is that as the number and difficulty of the tasks required of the driver

---

*Not a humanoid robot with human features, but a robotic replacement for the functions exercised by an experienced and safe human driver.

are reduced, the driver's attention may wander or may become occupied by unrelated tasks, with the risk being that the driver may not be able to respond sufficiently quickly to take control in circumstances where the robot driver cannot cope – such as sudden reduced visibility or a difficult-to-identify obstruction.

This problem of achieving a safe handover to the human driver has prompted the *revolutionary approach* to automation: in this approach, the vehicle is designed not to need a human driver at all. When these vehicles are purpose built, the steering wheel and other controls may be omitted; where a conventional vehicle is adapted, conventional controls normally become redundant. Waymo, a subsidiary of the parent company of Google, is probably the most advanced in developing vehicle automation, operating a driverless taxi service – a 'robotaxi' – in the East Valley of Phoenix, Arizona, without safety backup drivers in the vehicle. However, the road, traffic and weather conditions in that location are particularly favourable, and it remains to be seen to what extent such a service could be extended to more demanding situations. More generally, the concept is that of an autonomous vehicle able to function without a driver within a defined geographical area where conditions are suitable. It would be more demanding to go beyond that to replace the human driver under all conditions, although this is the ultimate objective of the proponents of the technology.

Waymo/Google has developed driverless technology because it has the expertise to do so and judges the potential as likely to be rewarding. If its efforts prove successful, it would make the technology available to car manufacturers but does not plan to manufacture vehicles itself. In this respect it resembles Google's sponsorship of the Android mobile operating system, which is available to all manufacturers of mobile phones. There are a number of other enterprises pursuing driverless technology, mostly involving tech start-ups supported either by car manufacturers (who see this as potentially an essential technology) or by deep-pocketed businesses that are seeking a commercial opportunity.

Benefits of automation

The real benefits of vehicle automation are less clear, in my view, than enthusiasts assume. Proponents argue that there are safety benefits, especially in the United States, where 36,000 people were killed in motor vehicle crashes in 2019. Given that human error and risky behaviour is responsible for 90% or more of fatalities, it would seem reasonable to expect that robots could do better than fallible drivers. On the other hand, robots suffer from their own shortcomings, tending to be less effective at perceptions involving high variability or alternative interpretations. In particular, robots would find it difficult to engage in the kind of visual negotiation that occurs between human drivers to settle which gives way when space is tight. Moreover, the driving performance of a robot would need to be very similar to that of a human driver to ensure public acceptability. A robotic driver that proceeded particularly cautiously to meet safety requirements would be unattractive to purchasers of AVs. The robot driver will therefore need to learn how to drive like a human.

Fatalities involving AVs, although rare, naturally attract attention. It seems likely the public will expect an AV to perform substantially better than a regular car, but exactly how much better is an important question. In any event, it will be difficult to demonstrate the safety performance of AVs in practice. For instance, in Britain there is one fatality per 140 million miles driven, so if AVs are to do better than a human driver, fatalities will be exceedingly rare events. Another issue is that achieving and demonstrating cybersecurity will be crucial for AVs since a hack of a fast-moving vehicle would be dangerous.

Those who urge high safety standards at the outset of AV deployment stress concerns about the public acceptability of the new technology. Against that is the argument that the best can be the enemy of the good, meaning that any automation that reduces crashes should be deployed as it becomes available, with the expectation that safety performance will improve over time. Automation should indeed improve safety over time, yet there are many other measures that would reduce deaths and injuries (e.g. the Vision Zero approach discussed earlier) that are almost certainly more cost-effective than vehicle automation.

Another claimed benefit is that automation might increase the capacity of existing roads by allowing vehicles to move with shorter headways, i.e. with a smaller distance between them than the recommended two-second gap on fast roads. The more precise control exercised by a robot might also smooth traffic flows and allow the use of narrower lanes. However, such increases in capacity seem likely to be possible only on roads dedicated to AVs since the presence of conventional vehicles, not to mention cyclists and pedestrians, would require standard spacing to be maintained. In any event, any increase in capacity would be expected to attract additional traffic (another instance of induced traffic), so congestion relief would not be expected. Automation that allowed an increase in road capacity might therefore be of interest to a road authority, but not to vehicle owners who would bear the cost of the necessary technology.

More generally, because AVs would be capable of operating empty, e.g. when returning to base after dropping off their occupant, they could add to traffic and hence to congestion. Conventional taxis operate without a passenger while seeking a fare, of course, but privately owned vehicles without occupants would be a new source of traffic and may require regulation if congestion is not to increase.

Prospects for automated vehicles

The prospects for widespread vehicle automation on existing roads with mixed traffic seem very uncertain. The driving task on motorways might be lessened in good visibility and in the absence of roadworks, but the driver would need to be immediately available to take control in adverse situations; one approach to this problem is to monitor driver attentiveness, e.g. by checking that hands are on the wheel and alerting the driver if attention wanders. There are low-speed environments that might accommodate AVs, including campuses, business parks and other planned developments with extensive road space. Yet it is hard to see robotic vehicles negotiating historic towns and cities with complex layouts, often narrow streets, extensive kerbside parked cars, cyclists and pedestrians.

This urban impediment to automation is particularly relevant to the robotic taxis that could be attractive to ride-hailing operators, to avoid the cost of the human driver and spread the extra capital cost of the robot through intensive use. On the other hand, there would be the potential cost of ownership of fleets of automated taxis, which contrasts with the asset-light business model of ride-hailing businesses where drivers own their own vehicles.

Commercial roll-out of robotaxis and privately owned AVs may need local permits to comply with the requirements of particular cities, which would limit the rate of deployment of the technology and the returns to investors. In contrast, the evolutionary approach to vehicle automation allows additional facilities to aid the driver to be offered across the whole vehicle fleet, subject to satisfying safety regulations, provided the attractions to purchasers justify the incremental cost. This is the usual way in which the auto industry has developed its products, so the evolutionary business model may be more attractive commercially than the revolutionary approach, even if the eventual outcome is not entirely driverless travel.[36]

After more than a decade of AV development, the early excitement and optimism have been followed by some disillusion as the problems of achieving an acceptably safe product have been recognised. Nevertheless, resources continue to be committed to refining the technology, which many see as a natural development of the 'automobile', a term whose etymology derives from the Greek *autos*, meaning 'self', and *mobilis*, meaning 'movable'. It would be premature to predict the eventual outcome – either the timing or the scope for wide deployment. It is certainly possible, as enthusiasts for the technology assert, that children born today will not need to learn to drive a car. However, it seems unlikely that automation would increase the average speed of travel on the existing road network, which is constrained by safety and congestion.

The promise of new technology

A key finding from this chapter is that none of the four major new transport technologies – electric propulsion, automation, digital platforms and digital navigation – will increase the speed of travel. This is in marked contrast to the transport technologies of the nineteenth and twentieth centuries, which gave us step-change increases in velocity and, therefore, access to more destinations with more opportunities and choices of all kinds. The new technologies are being taken up if they offer a better quality of journey and cause less harm to the environment and/or people's health. Ride-hailing and digital navigation are popular. Uptake of electric vehicles is happening partly in response to driver interest but mainly because of regulations that require manufacturers to reduce emissions of carbon and air pollutants. Uptake of automation will reflect the benefits perceived by drivers in relation to the costs and performance of the technology.

Aside from these four major new technologies, there are some others that may permit faster travel in smaller markets. The deployment of the already established high-speed rail technology, in the form of the HS2 project, is intended to reduce the travel time between London and Birmingham by 37 minutes, giving a significantly faster journey than is presently possible. Yet rail is responsible for only a minority of all journeys, and HS2 will account for a minority of a minority and will therefore have negligible impact on the average speed of travel.

Another niche market is urban aviation, where safety and noise concerns limit helicopter use. A number of aeronautical specialists are currently devising electric alternatives for aviation, taking advantage of battery improvements for electric road vehicles. Many versions employ multiple small rotors driven by electric motors, as for small unmanned drones. These are quieter and safer than a helicopter because failure of a single rotor can be managed. There are also electric aircraft with fixed wings in development. These will achieve a greater range than is possible with rotors, and they may be feasible for domestic routes. Generally, though, battery capacity is the limiting factor. The most likely market is city centre to airport journeys for passengers willing to pay a substantial premium over a regular taxi fare. However, the safety regulators will need to be satisfied, as will those concerned with environmental impacts, before a commercial service could be offered.

Electric propulsion is not yet feasible for most air travel. Other possibilities under investigation – with the aim of reducing carbon emissions – include biofuels of various kinds, although availability may be a limiting factor and costs are substantially higher than kerosene. Biofuels can be used with existing aircraft types, whereas hydrogen as a fuel would require new designs. If it is widely adopted for heating, ground transport and materials manufacture, hydrogen might become available at an acceptable cost for aviation, whether as a fuel for a jet engine or to feed fuel cells that generate electricity for motors to turn propellers. But these alternative aviation fuels will not affect the speed of travel anyway, and they will not therefore reduce the time it takes to get to a specific destination or increase the number of destinations that can be accessed.

All in all, the main impact of new transport technologies will be to reduce environmental harms, by replacing fossil fuels with electric propulsion. They will also improve the quality of our journeys but without having much effect on the amount of travel we undertake.

Chapter 5

How did the pandemic change travel?

The coronavirus pandemic caused a shock to national and international transport systems that was unparalleled outside times of war. The immediate consequences and responses are generally well known. What is more important, but as yet less clear, are the consequences for the longer term: will there be lasting changes in why and how we travel, and will the transport system develop differently as a result? An important question is whether we will be able to deal more effectively with the problems identified in previous chapters as we recover from the pandemic.

Immediate consequences

Figure 5.1 shows how car traffic in Britain was affected by the constraints placed on social behaviour – to slow the transmission of the coronavirus – that started in early 2020. Traffic was 70% lower in April 2020 compared with the pre-pandemic baseline, but it then gradually rose as services reopened, reaching a level only 10% below baseline by the autumn. A resurgence of the virus led to further restrictions being imposed at various times over the rest of the year, each of which was accompanied by a downwards shift in traffic volumes. Car traffic had reached pre-pandemic levels again by the summer of 2021, as the public took reassurance

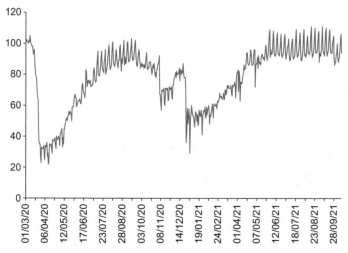

Figure 5.1. Car use as a percentage of the pre-pandemic level. (*Source*: Department for Transport Covid-19 transport use statistics.)

from the rollout of the vaccines, but there was then another dip in car use at the end of 2021 when working at home was encouraged during the Omicron phase of the pandemic.[1]

Public transport was also severely affected, with rail being hit worse than buses, as shown in figure 5.2 for London (similar outcomes were observed in other parts of Britain). Bus use fell to less than 20% of the normal pre-pandemic level in April 2020 before recovering in the autumn. London Underground use fell by 95%, as did national rail passenger numbers. Both bus and rail use fell again towards the end of 2020 as infections rose and social restrictions were imposed, but they then steadily recovered through 2021, with rail and bus use reaching 60% and 80% of normal levels, respectively, by the autumn. Public transport passenger numbers then dropped again at the end of 2021, during the surge of the Omicron variant. The huge falls in passenger numbers had a disastrous impact on the finances of public transport providers and the government had to provide very substantial funds to prevent insolvency.

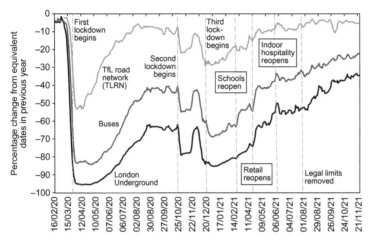

Figure 5.2. Travel in London as a percentage of the pre-pandemic level. (*Source*: TfL, Travel in London Report 14, 2021, figure 2.9.)

The onset of the pandemic stimulated initiatives by both national and local government to promote active travel (walking and cycling) as means of reducing crowding and virus transmission on public transport. The government issued formal guidance to local authorities that it expected them to make significant changes to their road layouts to give more space to cyclists and pedestrians, and substantial government funds were provided to local authorities for this purpose. As well as being an immediate response to the pandemic, the policy was intended to encourage alternatives to the car for short journeys as one element of the policy to decarbonise the transport system.

Many local authorities took advantage of the funding to quickly put in place 'low traffic neighbourhoods' (LTNs). These are areas in which car use is impeded by physical measures such as bollards in the roadway, while cyclists and walkers are free to travel, benefiting from the reduced general traffic – a technique known as filtered permeability. However, in a number of well-publicised instances, considerable local opposition arose, particularly from motorists who felt they were being disadvantaged. Resistance to traffic constraints also came from London black cab drivers who went to the High Court to challenge their exclusion from a major street, initially winning their case. The judge ruled that the plans of the London Mayor and TfL to rapidly repurpose streets were not justified by the pandemic. TfL appealed to a higher court, however, and the earlier judgement was overruled, essentially upholding TfL's approach as reasonable.[2]

The pandemic led to a large increase in cycling nationally (see figure 5.3), with peaks in recreational use at weekends. The general increase continued until the summer of 2020 but then declined in the autumn and winter (a result of poorer weather). Cycling fell to below the pre-pandemic baseline in early 2021, partly as a result of more people working from home rather than commuting to an office, and then reverted to pre-pandemic levels as 2021 progressed. With the benefit of hindsight, it would be hard to argue that what ended up happening to the level of

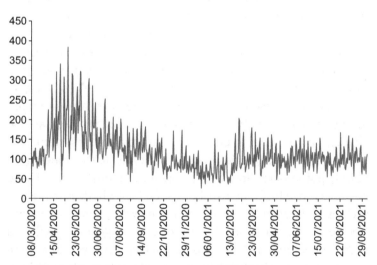

Figure 5.3. Cycling as a percentage of the pre-pandemic level.
(*Source*: Department for Transport.)

cycling justified the emergency reallocation of road space for that purpose, taking away capacity for general traffic.

There was a dramatic 98% fall in the number of passengers travelling through UK airports in April–May 2020 compared with the previous year. This was followed by a partial recovery during the summer as travel restrictions eased, but numbers were still down by 80% in 2020 compared with the previous summer. Subsequent travel restrictions meant that passenger numbers were down by nearly 90% in December, and for the year 2020 as a whole the reduction was 75%. By the summer of 2021 passenger numbers had still only reached approximately 20% of the pre-pandemic level.[3]

I mentioned the National Travel Survey in previous chapters, the findings of which have shown stability in average travel behaviour over the past two decades: there have been about 1,000 journeys per person per year, with an hour a day of travel time, covering some 6,500 miles per year by surface travel. As expected, the 2020 report showed a marked decline, with the average number of journeys falling to 739, with three-quarters of an hour per day of travel time and 4,334 miles covered over the course of the year. The question is whether and when we will revert to past travel patterns. This is the topic to which we now turn.

Longer-term consequences

As I finalise this book in February 2022, we are still actively dealing with the coronavirus pandemic in its Omicron variant phase, and we are facing considerable future uncertainty: about the possible emergence of new, even more infectious or more harmful variants; about the extent of protection offered by vaccines; about the population's level of immunity; about the promise of new medical treatments to counter the progress of viral infection and mitigate symptoms; and about how social behaviour will develop in response to a virus that will become endemic rather than disappear. Nevertheless, we may reasonably hope that we will achieve effective containment such that the future annual excess mortality attributed to Covid-19 is comparable to that of seasonal influenza, or not too much more.[4] We will learn to live with the virus, and this means that society will return to 'normal', but will this be the old normal or a new normal? In particular, what will the consequences be for travel and transport?

The starting point for consideration of the pandemic's longer-term consequences is the economists' concept of travel as a 'derived demand'. What is meant by this is that we travel in order to gain benefit from the destinations we reach. The value of these benefits must be sufficient to justify the time and money spent travelling.

The derived-demand perspective, while broadly valid, does not capture all the motivations for travelling. There are also what might be termed 'intrinsic benefits' to being on the move, irrespective of the destination: the benefits we experience particularly on recreational trips, whether walking, cycling, cruising on water or motoring on the open road. Our daily travel may include the intrinsic benefit of exercise when we walk or cycle, or of engagement with our local environment when we take an alternative route to a regular destination. These intrinsic benefits of mobility surely contribute to the finding that average travel time has remained unchanged, probably over centuries. Spending less time on travel would deprive us of the benefit of being on the move as well as the loss of benefit from accessing our chosen destinations, so I would not expect our desire to travel to reduce once the pandemic is behind us.

Time constrains our travel behaviour: there are only twenty-four hours in a day and lots of activities that must be fitted in. The impact of this unchanging time constraint has been overcome over the past couple of centuries by technologies that allow us to travel faster, and therefore further, gaining access to a wider range of people and places, opportunities and choices. As I discussed in chapter 2, we had reached a stable situation prior to the pandemic, in that the average distance travelled annually had stopped growing since the beginning of the present century. The question is whether the impact of the pandemic has disturbed this equilibrium in which travel costs and access benefits were in broad balance.

It seems unlikely to me that our desire for the benefits of access will reduce in the future. Yet the pandemic has shown that virtual access can substitute for physical access: when we meet online rather than face to face, for example. The trend towards virtual rather than physical access is at least partly reversing as the requirement for social distancing diminishes. Maintaining social distance comes unnaturally to humans who enjoy socialising, mingling, conversing, encountering, networking.

I also think that it is probable that the average amount of time we spend on the move is unlikely to change, given the long-term finding of the average hour a day spent on travel. So if we travel less for one purpose, we are likely to travel more for another. Accordingly, what may change are the *patterns* of travel: the particular modes we use. If we work part of the week at home and therefore travel less often to work, we may choose to move house to a more distant location and consequently make longer commuting trips. And if we do not move house, maybe we will make more local journeys, whether on foot or by bicycle or car.

Yet there is a substantial degree of stability to travel patterns – a consequence of the stability of the built environment in which the vast proportion of our journey origins and destinations are located. Changes happen quite slowly, and mostly at the margin. The number of new homes built each year adds less than 1% to the total housing stock, so a chain of house moves will generally involve existing properties and have little overall impact on the residents' demand for travel. The prosperity of cities can rise and decline, and populations ebb and flow as a consequence, and this does have an impact on the journeys made. The population of Britain is expected to grow in the future, although more slowly than it has in the recent past. The population is ageing, too, which has an effect on travel patterns.

These demographic changes are generally quite sluggish, but over time, slow changes add up. London, for instance, grew from 1 million inhabitants at the beginning of the nineteenth century to 8.6 million by 1940. As I described in chapter 2, there was then a fifty-year period of population decline as people sought a better life beyond a rundown, war-damaged city with polluted air, the population bottoming out at 6.8 million in the 1980s. But the tide turned in the 1990s as people were attracted back to a cleaner, thriving metropolis, and the population had reached 8.6 million again by 2015. Pre-pandemic, London's population was projected to grow to more than 10 million during the 2030s, and then to almost 11 million by 2040.[5] The pandemic might be expected to affect this growth, however, although the impact is hard to assess, because of both the temporary suspension of surveys to measure in and out migration and the impact on migration of restrictions on international travel.

Migration has also been affected by the state of the city's economy, which has been hard hit by the pandemic. Some temporary immigrants returned home while others worked remotely, as did some London residents who escaped to the country. There is considerable uncertainty about the overall position. One analysis suggests that the resident population of London may have been 700,000 lower in summer 2020 than a year before, but much of this reduction might have been temporary.[6] However, the judgement of London's official statisticians in May 2021 was that the scale of the fall in population is likely to be far short of the more dramatic figures reported.[7] Additionally, the Office for National Statistics estimated London's population in mid 2020 as 9.0 million – the highest ever, having grown by 40,000 in the preceding year, despite the pandemic.[8]

The urban dynamism that has been seen in London and many other cities around the world has been driven by the economics of agglomeration: larger labour markets, better supply chains, the knowhow that is 'in the air'. Agglomeration effects also operate for people's cultural and social lives, with more choices of places to go and people to meet, not least in the dating market, with opportunities broadened by apps that offer more nearby potential partners where population density is high.

Yet there are countervailing pressures to urban agglomeration, including high rents and house prices, travel congestion and overcrowding, and likely continued anxieties about infection by respiratory viruses. There may be specific issues for a particular cluster of business activities – something that is illustrated well by the heyday of Fleet Street, the physical home of national newspapers, with editorial staff on buildings' upper floors and printers in the basement. The compact location allowed staff to be shared, news to travel fast and gossip to flourish, but the downside was higher costs – both from transporting the printed product from central London around the entire country overnight and from restrictive labour practices. The advent of modern computing allowed journalists to input copy onto pages that could be printed at remote printworks that were better located for serving dispersed market outlets. The editorial offices then scattered to locations elsewhere in London. Nowadays, when someone talks about Fleet Street, they are usually referring to the British national newspapers collectively and to the journalists who work for them and not to the actual location. Indeed, during the pandemic, most journalists worked from home.

The relative stability of the built environment and the gradual nature of demographic change have both served to limit the impact of the coronavirus pandemic on travel behaviour. During the pandemic, anxieties about the risk of infection in densely populated urban areas as well as the new opportunities for remote working prompted some to think about moving out of the city into the countryside. This led to a shift in the relative prices of city and country properties, in a context in which house prices generally have tended to rise as people seek more living space to accommodate homeworking and because, overall, supply falls short of demand. Yet local resistance to house building on rural greenfield sites means that the prospects for additional country living will be limited. The main opportunities for new home construction are in existing urban areas, which have indeed experienced marked increases in population density in recent years.[9]

It seems unlikely that the coronavirus pandemic will change the longstanding attractions of city life, especially for young adults early in their careers and relationships. Throughout history, cities have largely recovered from plagues, as well as from physical destruction in times of war. An analysis of historical events that affected infrastructure demand – including the impact of the 2003 SARS outbreak on the use of the Taipei Underground and of the 9/11 New York City terrorist attacks – found only a short-term impact on willingness to travel by public transport, as long as there were no further events to suggest that the risk would persist over time.[10] However, these were relatively short-term shocks. Prolonged disruptions are more likely to cause long-term changes in behaviour as individuals and businesses find alternative ways of meeting their needs.

The extent and timing of recovery from Covid-19 is, therefore, less certain, particularly in respect of those who have been able to work at home (70% of people in some affluent London suburbs but less than 15% in some northern industrial towns[11]). The rate at which people return to their offices may be constrained by anxieties about crowding on public transport travelling into city centres (and also in lifts within tall buildings). The extent of the recovery of city centres may also depend on the extent to which there are long-run changes to working practices.

Changing patterns of work

A major consequence of the pandemic was homeworking for those for whom that was possible. This affected about half the workforce – very largely those in the service and knowledge sectors, for whom digital tools and online communication were a reasonably effective substitute for face-to-face engagement.

Indeed, meetings via platforms such as Zoom and Teams increased efficiency by saving travel time, as well as permitting greater attendance at meetings. I was able to participate in a weekly series of well-attended seminars organised by the Massachusetts Institute of Technology at lunchtime in Boston, teatime in London – something I could not have contemplated before the pandemic.

Experience during the pandemic has shown the feasibility of more flexible styles of working in which the working week is divided between home and office. Surveys by the Office for National Statistics found that while 27% of the workforce worked from home at some point during 2019, that figure increased to an average of 37% in 2020, with the proportion at any given time reflecting the pandemic restrictions then in force (as these lessened, homeworking declined). Homeworking varied by employment sector, too: it was highest for those working in information and communications, at 80%, followed by the professional and scientific sector, at 75%, with hospitality lowest, at 10%. Surveying future expectations in April 2021, 24% of businesses reported an intention to use homeworking as a permanent business model, although 28% were unsure.[12]

For businesses for which homeworking is feasible, there is an opportunity to save on office rents, while staff would save the cost, time and discomfort of commuting. Indeed, as remote working becomes widely accepted, people could live wherever they choose, working for employers wherever they are located. This would allow organisations to access the talents of a wider range of staff. Perhaps the 'death of distance' concept, first articulated twenty years ago, will at last be widely realised.[13]

Yet there are benefits to congregating together with fellow workers. The pull factors include the informal exchange of ideas, knowledge, news and gossip through serendipitous encounters; innovation through brainstorming; the social capital embodied in effective teams; the ability to induct new recruits into the culture of an organisation; and a reduction in operational risk when staff are on-site. Think of musicians, for example, who are able to practice at home to improve their individual skills but who need to come together with fellow players in an orchestra to perform in harmony.

Push factors that motivate people to go into the office include wanting to be part of a team, camaraderie, social life at the end of the working day, and domestic circumstances that are not conducive to homeworking. There is also FOMO (fear of missing out) from not being around when opportunities arise, not being seen by the boss when he or she is in the office, and not being recognised for one's contribution. This may be particularly relevant for women, who may appreciate the flexibility of working from home but whose efforts may not be fully appreciated by their superiors. Beyond paid work, students need to be on their campuses to make the most of a formative stage of their lives.

A key question in a competitive economy concerns the effectiveness and efficiency of working at home. It has been remarkable that the financial services sector, with support from the regulators, has been able to achieve almost complete remote working while maintaining data security. While every firm in the sector is constrained by the pandemic, the relative efficiency of homeworking is not an issue, but once offices can be safely reoccupied, it would not be at all surprising if reverting to doing business on the large trading floors proved more efficient. In a sector where trading is intense, margins fine, profits visible and bonuses motivating, the competitive edge from being present could be decisive. Moreover, in a highly regulated sector, firms may take the view that they are less exposed to the risk of failures of compliance if staff are supervised on company premises. However, there may be differences in working practices between US banks that expect staff to be in the office and their European counterparts that are willing to allow flexible working (one consequence of which will be the ability to economise on rented office space).[14] The wishes of employees who prefer to work from home for part of the week would have to be taken into account, given the need for employers to recruit and retain talented staff.

For the more creative parts of the service sector, hybrid working may be attractive. There is a trade-off to be struck between expectations of greater creativity when it comes to working on new projects in the office, whereas there might be greater creativity on existing tasks working at home.[15] A portion of the week might therefore be spent homeworking, with less interruption, while individuals might come into the workplace for two or three days, say, to liaise with colleagues.

But which days? Working at home on Monday and Friday seems natural, but it may give the impression of a long weekend of relative leisure, keeping on top of emails while also engaging with family responsibilities. Other 3–2 plans are possible: Monday, Tuesday and Thursday in the office, say, with Wednesday and Friday at home. Or perhaps a 2–3 pattern: getting the whole team in the office on Monday to plan tasks for the week ahead and then bringing everyone in again on Friday to debrief.

A further possibility is for people to work in local low-cost workplaces, avoiding both the commute and space constraints at home. Interactions with colleagues are likely to be limited, though, so it remains to be seen whether this way of working will prove attractive. Since a disadvantage with hybrid working is the reduced likelihood of fruitful informal interactions with colleagues, it is possible that there will be more out-of-office awaydays to forge team spirit.

The pandemic has provided a shock to office-based working practices and a large-scale social experiment that seems certain to have lasting effects. Working from home has had better outcomes than expected – a consequence of investments in new technology and reduced stigma of homeworking. We may therefore expect to see increased remote, flexible working without the constraint of place. Organisations will need to understand how jobs and tasks can be reconciled with employee preferences to ensure the work gets done.[16] The trade-offs are quite complex and depend on individuals and the nature of their work, as well as on the businesses they work for. Impacts affect productivity and well-being, creativity and relationships, sharing tacit knowledge and building trust in colleagues and clients. For many, neither attending the workplace for five days every week nor working at home for five days will strike the right balance.[17] A September 2021 survey of managers and staff found that 88% of employees wanted to keep working remotely at least some of the time, whereas 46% of managers had a more organisational concept of hybrid working that involved a mix with some staff working remotely all the time and some on site all the time.[18]

Where we will end up is not yet clear, as illustrated by the tension between two opposing views expressed at different times by James Gorman, the CEO of Morgan Stanley. First, he said:

> If you'd said three months ago that 90% of our employees will be working from home and the firm would be functioning fine, I'd say that is a test I'm not prepared to take because the downside of being wrong on that is massive.[19]

And then, in contrast, he subsequently stated:

> Make no mistake about it – we do our work inside Morgan Stanley offices, and that is where we teach, that's where our interns learn, that's where you build all the soft cues that go with building a successful career.[20]

While cities have generally recovered from plagues and wars, with life returning to the previous normal, it is conceivable that the kind of homeworking made possible by modern information and communications technologies will have a long-term impact on the urban working environment. A sampling of companies by the *Financial Times* in September 2021 found that while US banks were offering staff little workplace flexibility, most companies expected to adopt some form of hybrid working.[21] At present, it is difficult to judge the long-term outcome: either where we will work or how we will travel to and for work. That said, even during the pandemic there was competitive bidding for sites for new construction in central London, reflecting the expectation that modern, flexible, sustainable, high-quality offices with access to open air will continue to attract large corporations competing for talented staff.[22]

Demand by traditional clients for office accommodation in city centres would fall if homeworking were to persist on any significant scale after the pandemic, since those responsible for controlling costs in organisations would downsize and re-equip the workplace for shared activities. Such downsizing would tend to lead to a reduction in rents, which in turn would make space attractive for businesses that had previously seen these locations as too expensive, since well-located high-grade premises help attract high-quality staff. Good office space would therefore be unlikely to remain unoccupied for long, although the returns to landlords might decline. Outlets in office-heavy neighbourhoods that provide food and drink for office staff would continue to be in demand.[23]

Lower-quality office space might be repurposed or rebuilt, particularly for homes, although the depth of floors of existing buildings may limit their conversion to flats. Repurposing is nothing new. The conversion of old warehouses to loft apartments has proved popular. Virtually all the Georgian and Victorian terraces of inner London (in Bloomsbury, for instance) have been repurposed from their original residential use to accommodate diverse commercial, educational and cultural activities. They could revert if demand for such central workspace declined. One consequence would be more people living within walking distance of their work: the '15-minute city' concept.

How much difference would a shift to homeworking make to the transport system? It depends on both the amount of homeworking and the timing of people's working days. To the extent that people can choose when to work from home, avoiding road traffic congestion and crowding on public transport would be a factor that would tend to even out commuting across the working day and working week. That might also apply where work teams are able to decide collectively when to be in the workplace. On the other hand, when the total capacity of the workplace is substantially less than the total staff employed, businesses might decide to actively manage time spent in the workplace, hotel style, to even out the use of the available space across the week.

As things seem at present, there need be no substantial reduction in the numbers commuting to city centre employment in the long run. Yet there remain questions about whether we will get back to the previous density of people on buses and trains, and in lifts and offices, and if we do, it is unclear how long it will be until we are willing to be packed like sardines on commuter rail routes. Perhaps we never will revert to what, with hindsight, might be seen as absurd overcrowding. All planes and some long-distance trains are seating-only, and so are modern football stadia, where closely packed standing was permitted in the past (although there are current experiments to see if standing can be made safe). During the pandemic, buses limited the numbers that were allowed to board. We will see if there develops a demand from users of public transport for more civilised conditions and whether the operators are able to respond.

During the pandemic, the increase in the number of people working from home meant less commuting, although the saving in transport carbon emissions was offset by increased home heating. For those who had to get to work, car travel was an attractive alternative to public transport, so long as parking space was available at the place of work; walking and cycling were similarly attractive over short-to-moderate distances. Once the risk of infection is seen as quite low, I would largely expect people to revert to their previous modes.

The loss of passengers on public transport during the pandemic was financially crippling to the bus and rail operators, who had to be bailed out by the government. It is not yet clear to what extent fare income will recover. Any shortfall could lead to a reduction in services, which in turn could set in train a spiral of decline, a remedy for which would be increased public subsidy on the grounds that good public transport is essential for any successful city.

The pandemic put a halt to most business travel, with remote communications and video conferencing substituting, generally effectively. Given the cost of business travel, in terms of both time and money, some reduction compared with pre-pandemic levels might be expected. However, a survey of UK companies in January 2021 found that while 41% expected to make fewer business trips after the pandemic, 30% expected to make the same number and, surprisingly, 27% anticipated making more; businesses expected to be using a similar mix of travel modes as before the pandemic.[24]

Again, the question of competitiveness arises: would a client be more impressed if you committed the time and expense to meet face-to-face, or would they be happy to engage remotely? A likely expectation would be to meet new clients in person to secure a deal, but thereafter it might be acceptable for routine meetings to be conducted online. There has been a long-running debate as to whether new information and telecommunication technologies would substitute for physical travel or whether it would actually foster travel on account of the wider networks of contacts that can be maintained. The experience of the pandemic might shift the balance towards virtual rather than physical presence, particularly given the quality of the latest online communications technologies. Yet we should not underestimate the importance of personal engagement for establishing trust and commitment – just as important in an informal setting, over a drink or a meal, as in a formal meeting.

Air travel is the other mode of 'public transport', in the sense of transport that is available to the public at large who are willing to pay the fare for the trip. This sector was hit hard by the pandemic, but the expectation is that leisure travel will rebound in full to destinations that are sufficiently free from coronavirus. Business travel may not resume to the full previous extent, given the good experience of remote meetings during the pandemic. This would be damaging to the finances of airlines that cater to profitable business class travellers, although less so to the budget airlines. The incentive to increase airport capacity will be reduced and perhaps even eliminated, particularly the contentious third runway proposed at Heathrow.

Changing patterns of shopping

There has been a large increase in online shopping during the pandemic, partly so that people can avoid visiting crowded stores and partly because of the closure of non-essential shops during the most severe restrictions on social interaction. Businesses that could meet customer demand both online and in person generally did well, and those serving customers online only were able to thrive. Those operating mainly through high-rent city-centre premises, however, performed badly, and many long-established stores went out of business, particularly in major cities like London, Birmingham, Manchester and Glasgow. Retail in the small and medium-sized cities bounced back much more rapidly after restrictions were lifted, however.[25]

The shift to online shopping during the pandemic accentuated an existing trend whereby the average annual number of shopping trips in Britain had been declining steadily: from 228 per year in 2000 to 181 in 2019, and then falling to 141 in 2020.[26] The restrictions on non-essential retail imposed during the pandemic had a direct impact on sales in high streets and shopping centres, but online sales rose rapidly: they reached 60% above pre-pandemic levels during the second half of 2020. This constituted a marked uptick to a rising trend: online sales accounted for 5% of all retail sales by value in 2008, increasing steadily to 20% by 2019, and then rising to 30% in 2020.[27]

Yet shopping is also an enjoyable, and often a social, activity, and a rebound may therefore be expected for in-person retail, especially in city and town centres where other amenities attract visitors. It may be that the day of the department store is largely over, replaced by speciality shops and online retail. The upper stories of the city-centre stores may find new purposes, leaving the ground and first floors for fashion goods. This may result in some diminution in travel to shops in city centres, which might reduce crowding on public transport at peak times, although the new occupants of the repurposed upper floors will also have travel needs. The growth in online shopping has meant more van traffic delivering purchases, particularly on minor roads, as noted in chapter 4.

Hospitality and entertainment have been hard hit by the pandemic, but the underlying demand is unlikely to have been affected and bounce back is already occurring for those who are comfortable that large gatherings are safe. Indeed, to the extent that people will travel to work or for shopping less in the future, they will have more time for other things, including travel for other purposes, such as visiting family and friends or for leisure trips. If the past experience that we travel for about an hour a day continues to hold, then fewer journeys of one kind will mean more of another, with little change in the overall amount of travel, although the pattern of trips may change. Some of this travel may be purely recreational, and some of that may not be counted as 'travel' by the statisticians who compile the data on how much we travel, e.g. they disregard walking in the local park; but if we get in the car to go out for a country walk some distance away, that counts as a leisure trip.

Necessary journeys

The response to the question posed by the wartime poster that illustrates the introduction to this book – 'Is your journey really necessary?' – comes in two parts. During the pandemic, we found we could manage with a lot less travel. But as the pandemic has receded, we have found that much of the travel we undertook before was important to how we live, and the signs are that we wish for normal service to be resumed for many purposes. On the other hand, the restrictions of the pandemic lasted long enough for some new habits to be formed – particularly online shopping and working from home – which are changing travel behaviours in the longer term, to some degree. Although the pandemic has shown that rapid and substantial changes to travel behaviour are possible, it seems unlikely that we will be willing to lose the access benefits that travel provides.

The potential personal impact of coronavirus infection created a very strong incentive to reduce the amount we travelled. In contrast, although we generally recognise the imperative of tackling climate change, the timescale of this emergency spans decades and the impact of changing our individual behaviour is minute, so we will be reluctant to forgo the benefits of travel. In short, much of our travel is not really necessary, and if pressed we can manage with less, but travel is what we want, so the 'new normal' is not likely to be very different from the old.

The impact of the pandemic on the transport system constitutes what is known as a natural experiment: a major change effected by a natural event much larger than could be achieved by a deliberate intervention. The experiment is only part way through. The long-term impact of Covid-19 is not yet clear. My hypothesis is that lasting changes in travel behaviour as a result of the pandemic will be limited, and that we will therefore broadly revert to how we travelled in the previous twenty years: on the move for about an hour a day on average, with about a thousand journeys being taken each year, and the distance travelled being about 6,500 miles per year by all surface modes. This is the consequence of a balance between the access benefits of travel and the costs of time and money, together with a built environment that determines origins and destinations of journeys and that changes quite slowly.

Subject to these constraints, there could be some changes in travel behaviour:

- less commuting and fewer visits to shops, replaced by online access;
- less business travel, replaced by virtual meetings;
- less use of public transport for traveling to work and city centre retail; and
- more leisure trips, to make up for travel time saved for other purposes.

The extent and timing of such changes are not yet clear, so we cannot count on post-pandemic travel behaviour being sufficiently different to pre-pandemic patterns to ease the task of tackling the problems of the transport system, as I discuss in the next chapter.

The key conclusion of this chapter is that experience of the pandemic, in which overall demand for travel was substantially cut, does not point to evident opportunities for reducing car ownership and use.

Chapter 6

Building back better?

There has been widespread hope that the experience of the coronavirus pandemic will lead to certain things getting better as we move forward: that is, that the new normal will be better than the old. This chapter will try to identify how travel and transport could be improved in the light of this experience.

One indication of what is possible is the way that science, technology and public service have been harnessed to develop and deploy vaccines in record time. Might such energy and urgency be applied to other societal problems, including those that are encompassed by the idea of sustainable development, and in particular the mitigation of climate change?

In the previous chapter I argued that the pandemic is unlikely to lead to substantial changes in our travel patterns, once society has adapted to the new endemic coronavirus. The amount of travel before the pandemic reflected a broad equilibrium between the benefits of access to destinations and the costs of travel (in terms of both time and money). The destinations are very largely located within the existing built environment, which changes quite slowly. Given the distances between established destinations, the car has become the main mode of travel to allow the level of access to which we have become accustomed. And while there is scope for virtual access substituting for physical access – whether that is for work or for shopping – the travel saved is quite likely to be replaced by travel for other purposes.

This means that the transport-system problems that we faced before the pandemic remain as those we must tackle after it. One difference is that we know we can manage with less travel if there is a pressing reason to do so. Yet while the urgency of the pandemic, with its direct consequences for our health, was a strong motivator to change our travel patterns, the climate emergency is chronic rather than acute: for most of us, the personal impacts seem rather distant. Although changes to our individual travel choices would have only a minuscule impact on carbon emissions, those who are most alert to the progress of global warming try to act in a way that demonstrates what is possible for the rest of us. Nevertheless, changing why and how we travel to respond to a changing climate is a bigger ask than responding to a pandemic, which is why many people argue that faster technological innovation is a more productive approach.

To consider how the transport system might be improved – both to better meet our travel needs and to mitigate its impact on the environment – I will discuss interventions at four levels: local, regional, national and global. Note, though, that decisions made at any higher level affect the levels below.

Local decisions

For neighbourhoods, the main issue is generally the impact of the car, both on the local environment (particularly air pollution and noise) and on the interactions between people. Roads and streets in cities, towns, suburbs and villages have two functions: as places where people do business and engage with one another, and for moving from one place to another. Roads that are heavily used for movement are less suitable as places for engagement. Transport for London has adopted a three-by-three matrix to generate nine categories of street, ranging from the arterial road, with maximum movement and minimum 'place function', to the 'city place', which accommodates limited vehicle movement but encourages a high level of pedestrian activity.[1]

There is scope for changing the balance between movement and place through physical interventions. In the last century, when car ownership and use was growing, more space was created for vehicles through the creation of new urban motor roads, sometimes elevated above street level, as well as by a variety of traffic management measures. In recent years, the damage caused to the urban environment by the resulting traffic flows has been widely recognised, and the pendulum has swung towards creating more space for pedestrians and cyclists at the expense of space for cars, either when moving or parked.

There is insufficient street space in densely populated urban areas for all the movement and place activities that might take place. When considering how to manage the competition for space, the different demands of the distinct modes of travel need to be taken into account. Compared with walking, a bicycle takes up about twice as much space, on average, and a car five times as much. In contrast, a bus can be more space-efficient than walking when it is fully loaded. On-street parking typically represents 20–30% of urban road space.[2]

Low Traffic Neighbourhoods

The Low Traffic Neighbourhood (LTN) is an approach that aims to limit through car traffic ('rat running', as it is known) while still allowing residents and their visitors, delivery vehicles and taxis to gain access to every address. The reduction in car traffic fosters active travel as well as allowing children and others to make more use of streets safely. The introduction of LTNs can be controversial on account of what is known as 'loss aversion': the tendency to prefer avoiding known losses to acquiring promised gains.[*] Shops in particular are often anxious about loss of

---

[*]It is said that the nineteenth-century prime minister Lord Salisbury, when approached by an assistant about the need for change, responded: 'Change? Change? Aren't things bad enough as they are?'

business from car users, failing to anticipate the increased footfall that may arise by making an area a more agreeable place for people to shop.

A November 2020 survey of the general public indicated high levels of support for government action that sought to increase road safety, improve air quality, and reduce road traffic, congestion and traffic noise in neighbourhoods.[3] A further survey of residents in LTNs in January 2021 found similar high levels of support.[4] In contrast, the London Borough of Ealing sought views on a number of LTN schemes that had been implemented during the pandemic and found that a majority of residents who responded opposed most of the schemes.[5] However, these were not representative sample surveys and it is probable that those who perceived a loss would be more likely to respond than those who perceived a gain. Despite this, after a change in political leadership Ealing Council decided to remove all but one of the LTNs it had implemented.[6]

Overall, the evidence suggests that the impact of LTNs is largely positive and that it is only in the longer term that most of the benefits become apparent.[7] However, their implementation can be contentious, and skilled handling by local politicians and planners is required. The introduction of experimental LTNs without the usual consultations – prompted by the Covid-19 pandemic – generated a lot of pushback from those who thought themselves disadvantaged. On the other hand, schemes to restrain traffic around schools during the times of day when children are arriving and leaving are generally supported, as are closures of streets for a few hours a week to create 'play streets'.

One particular concern is that traffic may simply be diverted to the roads surrounding an LTN. This can be portrayed as better-off people who own nice homes in a given neighbourhood ridding themselves of traffic that then flows past those living on nearby main roads. However, while there may initially be such diversion, experience tells us that for schemes that have been in place for longer, reductions in traffic are observed on most surrounding streets.[8] More generally, the evidence suggests that while constraining through traffic along one road leads to diversion to adjacent roads, an area-wide constraint results in the 'disappearance' of traffic.[9] This disappearance is a consequence of the time constraint on travel behaviour and how, therefore, congestion is self-regulating, as discussed in previous chapters. In brief, through traffic that is displaced to adjacent main roads increases congestion and delays, which results in some road users making alternative choices of routes, times and modes of travel. LTNs are most acceptable in cities where good alternatives to car travel exist.

Improving urban living

LTNs improve air quality and reduce traffic noise, accidents and air pollution within neighbourhoods, but to make wider improvements to air quality, many UK cities are adopting Clean Air Zones (CAZs), which levy a charge for entry on older, polluting vehicles. This can be seen as a temporary expedient until the uptake of zero-emission vehicles is sufficient to reduce NOx and particulate levels below the maximums permitted. London's CAZ is known as the Ultra Low Emission Zone (ULEZ) and its implementation had reduced NOx concentrations at roadside sites within the zone by 44% ten months after initiation.[10] The ULEZ was extended to the area within the North and South Circular Roads from October 2021, with a daily charge of £12.50 for polluting cars and vans. Birmingham introduced a CAZ in 2021, which, like London's scheme, covers older cars. Other cities are introducing similar zones but choosing not to charge private cars because the required air quality can be achieved by charging only larger vehicle types.

The most effective measure for reducing car use in the centres of towns and cities is to limit parking provision, both through higher charges and by prohibiting on-street parking. The most sophisticated approach to allocating on-street parking is dynamic pricing, as used in San Francisco, which adjusts parking charges regularly by trial and error based on parking space occupancy, with the aim of keeping about 15% of spaces available on each city block. Vacancies are notified to drivers on a smartphone app, which reduces the time spent and the pollution created from cruising while searching for a space.

More generally, there is a shortage of road space not only for moving vehicles but also for stationary ones, and not just for parked cars but also for taxis setting down passengers, for delivery vans in the process of unloading and for rental bike and e-scooter parking. Some cities have adopted policies relating to the allocation of kerbside space, with parked cars being given the lowest priority.

The balance struck between the space available for vehicles and that available for people is important for determining the characteristics of a neighbourhood. An idea that has recently attracted attention is that of the '15-minute city' (popularised by Paris mayor Anne Hidalgo) or the '20-minute neighbourhood' (a Melbourne initiative): a community in which essential amenities and services are accessible within 15–20 minutes of walking or cycling, thereby dispensing with the need for the car as the main means for getting around.[11] High-density cities give rise to smaller catchment areas for shops, schools and other facilities that require a sufficient number of clients to be economically viable. On the other hand, the scope for providing a substantial proportion of employment within such a tight area seems limited (and likewise for educational and cultural opportunities) without reverting to the equivalent of rural village life, with its quite restricted range of possibilities. Indeed, the reason for the historic trend of people moving into cities was to experience a wider range of opportunities, in terms of both the jobs and the amenities available within an acceptable travel time.[12]

In any event, there are limited possibilities for adapting existing urban environments for the new purposes required to create a 15-minute city, given the existing built environment and real estate ownership. Only half the trips made by Londoners take fifteen minutes or less, while prior to the pandemic only 17% of work-related trips made by active travel or public transport were undertaken in that amount of time.[13]

On the other hand, cities are not static, and a forward-looking town planning regime could encourage changes in use that reduce the need for motorised travel. For instance, permitting the repurposing of unneeded office space for residential use (in response to the increase in homeworking, as discussed in the previous chapter) may allow more people to live within walking distance of their workplace. It has been said that most of us spend time in two 15-minute villages: the one in which we live and the one in which we work.[14] What is not clear at present is the extent to which these two villages might merge in future.

Decisions about allocation of street space between 'place' and 'movement' – and between the contending modes of movement – are best taken locally since the most suitable options will depend on local circumstances and inclinations. Options to push back car use will depend in part on the availability of public transport (a matter that is addressed at regional level: see below) and in part on the extent to which cycling and walking can substitute for driving.

What's the use of cycling?

Many people are enthusiasts for cycling (I am myself, in moderation), and this includes British transport planners, who collectively rank walking and cycling as being top of their policy priorities.[15] Cycling accounts for just 2% of trips in Britain, so there is evidently considerable scope for an increase here, and that would generally be thought to be a good thing. But is it the solution to our problems with the transport system? If the answer is 'more cycling', what precisely is the question?

Consider the city of Copenhagen, famous for its cycling. It has excellent cycling infrastructure, with every main road having segregated cycle lanes. The overall share of journeys by bike in Copenhagen in 2018 was 28%, compared with 2.5% in London.[16] So when, at the beginning of the pandemic, London's mayor Sadiq Khan said he wanted to increase cycling tenfold, you can see that this should be possible, albeit with substantial investment in cycling facilities. However, when you are in Copenhagen, you realise that there is a considerable amount of general traffic – indeed, the car is responsible for 32% of trips, a figure only slightly lower than London's 35%.[*]

Aside from cycling, the other big difference is that public transport accounts for only 19% of trips in the Danish capital, whereas the figure for London is 36%. The evidence from Copenhagen therefore suggests that it is possible to get people off buses onto bikes, very likely because cycling is cheaper, healthier, more environmentally friendly, and no slower in congested traffic. But do we want to attract people off the buses, thereby reducing passenger numbers and fare revenue, which would in turn reduce service frequency or require higher subsidy? Moreover, buses are an efficient means of moving people around in urban areas, and they are capable of decarbonisation as zero-emission propulsion is introduced (powered either by electric batteries or by hydrogen fuel cells). More cycling is therefore unlikely to make much of a contribution to solving the problem of climate change unless drivers can be attracted out of their cars and onto bikes – something that seems difficult to achieve, even in a small, flat city with excellent cycling facilities like Copenhagen, where almost everyone has experience of safe cycling.

The available data for other European cities indicate that the picture in Amsterdam is similar to that in Copenhagen, with 32% of trips being by bike and 17% by public transport. In marked

---

[*]I recently rewatched a Scandi noir TV drama series set in Copenhagen and noticed that only one character used a bicycle; the rest rushed around in cars, as those in such shows always do, regardless of location.

contrast, both Zurich and Vienna have excellent public transport, trams in particular, which are responsible for 40% of trips, with cycling accounting for only 7–8%.[17,18] More generally, while the pattern of urban travel reflects both local geography and history, we do not find cities in developed economies that have high mode shares of both cycling and public transport.

The attractions of the car

It would be illuminating to understand why so many people opt for the car even in cycle-friendly cities. In the absence of evidence from surveys, I draw upon anecdote. Cars are useful for carrying people and goods, including child seats and the stuff you regularly use, as well as for making trips that are longer than would be comfortable by bike, enabling you to reach your destination with less sweat involved. The car is well suited for meeting our needs for access to people and places, including for trips with a chain of destinations, for door-to-door travel where there is road space to drive without unacceptable congestion delays and the ability to park at both ends of the journey.

But there is more to car ownership than the ability to go from A to B. For instance, there has been a noteworthy growth in the popularity of SUVs: from 10% of all car sales in Europe in 2010 to 36% in 2019.[19] While there are some practical reasons for preferring a car with large capacity and a high driving position, the rapid growth in use suggests that there are also feel-good factors that motivate the purchase of these costly vehicles (factors that are presumably known to the market researchers employed by the motorcar manufacturers but not generally published). Analysis of new car registrations in Britain found that large SUVs are most popular not in remote farming regions, where a 4x4 might have practical advantages, but in affluent urban areas.[20]

More generally, a relevant factor influencing car ownership has been termed 'car pride', by which is meant the personal image and social status associated with the decisions to purchase and use a car. A distinction can be made between two aspects of pride: feelings of self-worth and feelings of superiority in relation to others. These have been explored through social surveys to identify differences between cities and countries. The lowest scores for car pride were found in Japan, with most Continental European countries coming next; the United Kingdom placed rather higher, and the United States scored highest among the developed economies. The highest scores of all were those for India and other developing countries, where ownership levels are relatively low and the corresponding social status relatively high.[21]

Car pride contrasts with the notion of 'car dependency' – a term that is useful in two contexts. If the only practical way to travel is by car, e.g. if one lives in new housing built on a greenfield site with no public transport provision, then 'car dependence' is an appropriate way to describe the objective situation. In contrast, the term may be used – by implication with disapproval, perhaps analogous to 'drug dependence' or 'alcohol dependence' – to describe a situation in which there are seemingly better alternatives to the car that people choose not to use. Such a subjective judgement is likely to be contested by motorists, who generally find their cars to be liberating, permitting journeys that would be less convenient or less satisfactory by public transport, walking or cycling. This would include the motorists in Copenhagen who have the capability and opportunity to cycle but chose not to do so, for what we must suppose are, for them, good reasons. The notion of car dependency is found useful by those who aim to promote sustainable travel. They are concerned that their interventions are ineffective in reducing the demand for private motorised travel and that more radical policies are difficult to implement because of public and political acceptability concerns. They nevertheless recognise, reluctantly, that the car is deeply ingrained in our societies with consequent social barriers to sustainable transport.[22]

The fact that cars are generally parked for 95% of their lives is a good economic argument for the various forms of car sharing that are available. But conversely, this fact also indicates the value we place on individual ownership, to have a vehicle available when we want it – a vehicle that reflects our personal consumer preferences. Cars are not unique in this respect: many items that are found within a household are infrequently used, but we prefer ownership to rental.

We tend to underestimate the non-travel value of car ownership, and hence we underestimate this barrier to decreasing car ownership and use. The car is a great invention, the main attraction of which (as I discussed earlier) is the access to people and places it makes possible, with the opportunities and choices that ensue. It would not be easy to provide alternative means of access for much of the UK population, given the existing built environment and the prevailing patterns of travel that make such access possible. Active travel (walking and cycling) has many virtues – healthy exercise, low environmental impact, contributing to pleasant places – yet promoting cycling as an alternative to the car is an uphill task, and all the more so for walking, which is the slowest transport mode and therefore the least effective in achieving the access to which we have become accustomed.

Evidence of the difficulty of achieving a switch away from car use comes from the substantial UK government funding of local sustainable transport projects, worth £540 million over the 2011–15 period, which resulted in only quite small changes in travel behaviour: per capita car volumes declined by 2.6% for the project areas, compared with a decline of 0.3% elsewhere; cycling increased by 2.8%, compared with a decrease of 3.8% elsewhere; and carbon emissions may have been reduced by 1.5–3% more than would otherwise have been the case.[23] Another government programme worth £190 million over the 2013–18 period to fund new cycling infrastructure in eight cities found that most users were existing cyclists, with only 5% on average saying that they would have travelled by car had the improvements not been made.[24] These shifts in travel behaviour were quite modest given the effort and cash expended.

Generally, a change in behaviour occurs only when a person has the capability and the opportunity to engage in that new behaviour and is more motivated to adopt it than any other behaviour.[25] Getting people out of their cars and onto bikes therefore requires them to have the capability and the opportunity to ride on safe cycleways, and it requires them to prefer this mode of travel to taking their car. The key question in this context is motivation, in respect of which our understanding is insufficient.

Regional decisions

A good system of public transport offers an alternative to the car, particularly in urban areas where road traffic congestion and parking availability limit car use. Arrangements for public transport need to be made at regional level, within a national policy framework.

Rail in all its forms is faster and more reliable than the car on congested roads. Because it does not provide door-to-door travel, rail generally incorporates healthy walking into the daily commute. Buses are limited in speed and reliability by general traffic, but when they can be segregated from cars, they become more attractive to users. Priority bus lanes on regular roads are helpful. Fully segregated Bus Rapid Transit, as it is known, performs similarly to urban light rail but at lower cost.

As well as speed and reliability, the other important aspect of public transport is integration across the range of services (or lack thereof). One of the most admired public transport systems is that of Switzerland, which has the best timetable integration, with excellent connections between buses and trains and good coverage even for most rural areas. The country's 'Taktfahrplan' (Takt meaning musical beat; Fahrplan meaning timetable) provides a regular hourly rhythm of services at fifteen-, thirty- or sixty-minute intervals, with interchange hubs analogous to those employed by airlines. Such a system of public transport has the best chance of attracting those who would otherwise use a car.[26] Nevertheless, car ownership in Switzerland, at 500 per thousand inhabitants, is similar to that of neighbouring countries.

Better buses

Historically, buses in Britain were largely operated by municipalities. This changed in the 1980s when Margaret Thatcher's government privatised bus operations outside London with the aim of promoting competition, intended to drive down costs and better meet the needs of passengers, albeit with the loss of service integration. However, competition generally failed to materialise since bus operators found on-route rivalry yielded insufficient returns and preferred in effect to carve up the market, generally leading to a single dominant provider in each town. Overall, bus passenger numbers have been in decline where there is private sector provision, although there are some towns in which cooperation between the bus operator and the local authority has bucked the trend, Brighton, Reading and Nottingham being examples.

This general decline in passenger numbers contrasts with the picture in London, where Transport for London (TfL) retained overall responsibility for managing an integrated bus network, with private sector companies providing bus services under contracts that rewarded them for reliability. Bus use in London grew steadily until around 2010, when 4 million journeys were made each day; the total fell back a little after that, maybe partly due to road traffic congestion since demand for rail travel remained steady. TfL is also responsible for rail transport provision: Underground, Overground and Dockland Light Railway. TfL operates the first of these with its own staff and the latter two using contractors. The private sector businesses that provide bus and rail services in London compete from time to time to operate services on specified routes, which provides a check on efficiency and value for money.

Having responsibility for all public transport in London allows TfL to offer an integrated service, with a single cashless payment system that includes a capped daily charge, an online journey planner and transferable 'hopper tickets' for buses. TfL is generally regarded as a world leading public transport operator, despite having no subsidy from central government. Around 70% of TfL's income came from fares before the pandemic, which contrasts with much lower proportions for the transport authorities of other major cities: around 50% for Madrid; 40% for New York City, Paris and Hong Kong; and only 20% for Singapore.[27] A number of German and Austrian cities also operate integrated, region-wide systems of public transport, known as Verkehrsverbund, the creation of which has led to increased passenger numbers.[28]

London's success has prompted other UK cities to consider moving to an integrated system in which private operators provide vehicles to a city-owned public sector business that takes responsibility for the provision of routes, finances and planning generally. The government's new National Bus Strategy, mentioned in chapter 1, aims to make bus services around the country more like those in London, either by means of a city or region exercising overall control, as in London, or though partnerships between private sector operators and local authorities. Manchester is in the lead in introducing arrangements whereby the regional authority will assume overall responsibility to set bus routes, timetables and fares, with private sector operators providing bus services for payment dependent on achieving good quality of service.

More effective integration of bus and rail services, with good, well-understood levels of frequency, is the best way to encourage people to leave their cars at home. But experience has shown that good public transport with acceptable fares cannot be provided by private operators or public authorities wholly dependent on income from fares. Support from the public purse is needed, whether it comes from central or local government or from an arrangement analogous to the French 'versement transport', a regional payroll tax levied on the total salaries of all employees of companies with more than eleven staff, typically at a rate of 1–2%. This tax was originally intended to raise capital for investment in local public transport infra- structure but is increasingly used to cover operating expenses.

One way of improving bus services – both their speed and their reliability – is to reduce the competition for road space from general traffic. This can be done by physical means, e.g. by creating and enforcing bus lanes, or by limiting kerbside park- ing on bus routes (known as 'red routes' in London). A more general approach involves charging general traffic for use of road space, as in London's Congestion Charge zone. This was introduced in a zone in the centre of the city in 2003, and the volume of traffic in the zone went down by a third. There was an initial reduction in delays due to congestion, although this benefit declined over time, reverting to the previous level after five years. Nevertheless, the congestion charge has proved publicly acceptable and has functioned reliably, generating a useful amount of net revenue that is used to support transport services within London.

London's system levies a daily charge for entry into the Con- gestion Charge zone, and the charge does not vary according to the amount of congestion, the time of day or the precise loca- tion. There is a case for a more refined arrangement that would better reflect the level of congestion, with the aim of using the price mechanism to discourage driving at more congested times. There is also a case for extending the congestion charging zone beyond the present central area. However, these possibilities bear upon the question of a national scheme of road pricing to raise revenue from electric vehicles, which do not pay tax on petrol or diesel fuel, as we will discuss later.

Regional railways

Rail services are attractive for their speed and reliability, at least when they are well managed with adequate investment in trains and infrastructure. Cities are at an advantage if they have historic rail or tram tracks than can be modernised. Building new rail routes in urban areas is generally difficult. Underground rail is very costly and can be justified only in high-density cities, while space for surface rail is hard to find. Getting better use out of existing or abandoned railways can be worthwhile, as exemplified by the London Overground, which involved stitching together previously underused rail routes in inner London, with new rolling stock, renovated stations, high-frequency services and a distinct brand. Modern digital signalling allows intensively used rail routes to be operated with shorter headways (the safe distance between trains) and thus increased passenger capacity; for example, capacity on London's older sub-surface Underground lines is a third higher now than before signalling was upgraded.

Local, regional and national rail services overlap when commuters and other locals use long-distance services and stations for short trips. There is ongoing debate about whether the funds available for rail investment should focus on the regional or the long distance. The case for HS2 is about the benefits of improving the connections between London and the cities of the Midlands and the North, yet the National Infrastructure Commission has estimated that there would be bigger improvements to the economic and social lives of those cities from improving regional rail links.[29]

For road investment in the regions, the main question is how any additional capacity is used. We have seen that addition to the capacity of the Strategic Road Network can be largely taken up by local users, pre-empting the benefits to longer-distance business users for whom this network is primarily intended. Additional capacity on local roads can enlarge the catchment areas of businesses, in terms of both accessing skilled employees and reaching more customers, but the additional traffic thus generated tends to restore congestion to the same level it was before, negating the hoped-for benefits.

Britain has a mature transport system that is well used and needs to be well maintained. But the benefits from adding capacity are limited. The most attractive opportunities for investment are found in urban rail, which can increase access to dynamic city centres to enhance the economic, cultural and social benefits of agglomeration. However, such investment should be focused on those cities where strong economic growth is at risk of being held back by a transport system that cannot keep up with increasing travel demand. For most cities, journey times into the city centre at peak times are relatively short, meaning that further infrastructure investment is unlikely to improve economic performance.[30]

National decisions

For large local expenditures such as rail route improvements, the only source of funding is a grant from central government, which means that such investment decisions are made by government ministers. Substantial responsibilities are, however, devolved to the Scottish, Welsh and Northern Ireland governments, and many policy decisions made by the Department for Transport in London accordingly apply only to England, whose population amounts to 85% of the population of the United Kingdom.

A frequent demand is for a 'national integrated transport strategy', but the debate this has prompted has proved inconclusive as far as England is concerned. Scotland and Wales have their own national transport strategies, and London's mayor has a transport strategy as well. An important consideration in all three cases is that there are responsibilities and corresponding plans for land use, economic and demographic development, and sustainable growth; the transport strategy then fits naturally into place since the demand for travel depends on this wider context. However, the national planning policy frame-work for England is too broad brush to be the basis for a national transport strategy, and regional bodies, other than London, lack the relevant responsibilities. An integrated transport strategy for England is therefore a non-starter with the arrangements current in place for regional governance. Were there to be more devolution of responsibilities and budgets to regions, they could follow London's example, where TfL is a world-leading authority for both transport planning and operations.

Where to invest?

Britain's railways were transferred to private ownership in the 1990s, and the track then reverted to public ownership in 2002. However, the achievements of the private sector train companies that were awarded franchises were quite variable, with poor performance giving rise to public discontent. In 2018 the government set up a review under Keith Williams, a former chief executive of British Airways, with the aim of putting customers first. The much-delayed outcome of this review was published in May 2021 (as mentioned in chapter 1), and its primary recommendation was to recreate a unified rail system for Great Britain.

The main question for the future of rail is how much to invest in new capacity and where to make such investments. There has been remarkable growth in demand for rail travel in recent decades, with passenger numbers almost tripling between the low point of 630 million in 1982 and the high of 1.75 billion immediately prior to the pandemic. Investment in new trains and improved infrastructure has been a pull factor, as has the ability to work online while travelling. There were also push factors: congestion on the roads and the shift within the economy from manufacturing to business services that tend to be located in city centres, for which rail travel is well suited.

It is possible that this growth in rail travel has reached its natural limit, although the impact of the pandemic has made longer-term trends hard to discern. There are broadly two approaches to decisions about rail investment: either one can follow economic growth or one can attempt to lead it. The growth in London's population in recent decades, and the resulting crowding on public transport, has justified substantial investment in rail: new signalling on existing lines, allowing more frequent services using more rolling stock; and new routes, with the latest addition, Crossrail (which will be named the Elizabeth Line when it is opened), due to add an additional 10% to central London's rail capacity when it starts operating in 2022.

The question of causality is crucial to decisions about major transport investment, both rail and road: can new capacity *promote* economic growth, or is it needed to *permit* continued growth that arises for substantially unrelated reasons, as in London? Proponents of more investment in the Midlands and the North of England argue that this would stimulate economic dynamism, which is easy for them to argue since they are bidding for 'free money' from central government. If regions were financially autonomous, with devolved budgets and substantial tax revenues of their own, they would need to decide about relative priorities between transport schemes, broadband connectivity, environmental improvements, property development for inward investment, skills training and the rest. Good evidence for the impact of transport on local employment is very limited in extent and is decidedly mixed in terms of its conclusions.[31] Decades of research have not been able to pin down the causal relationships between transport investment and increases in economic performance as effectively as policy makers might like.[32]

I discussed investment in the Strategic Road Network in chapter 3. The questionable impact on economic growth certainly applies to the costly addition of incremental capacity to this mature network. Indeed, the findings of a study of eighty-nine US metropolitan regions were that a region's economy is not significantly impacted by traffic congestion, which is the main reason usually given for investment in road capacity.[33]

There are, in addition, particular problems with investment in road capacity, not least the use of additional capacity by local users, commuters and others, so negating the economic benefits intended for longer-distance business users. Another noteworthy drawback is public concern about the conversion of the hard shoulders of motorways to running lanes, which increases capacity without requiring further land take but can be the cause of crashes when a vehicle breaks down.

Yet the main problem with road investment that increases capacity for traffic is the additional carbon emissions created, which contribute to climate change. In the long run, electrification of the whole road vehicle fleet (or other means of decarbonisation) could deal with this, but in the near to medium term, most vehicles will continue to rely on oil fuels. How to decarbonise the transport system is the pre-eminent question for national decisions in a global context. This is the topic of the next chapter.

'Levelling up'

The idea of levelling up became a theme of UK government policy under Prime Minister Boris Johnson. A White Paper published in February 2022 set out the approach to remedying persistent regional inequalities. Twelve medium-term 'missions' were identified, one of which concerns transport: 'By 2030, local public transport connectivity across the country will be significantly closer to the standards of London, with improved services, simpler fares and integrated ticketing.'[34]

Although this mission is rather vague in its intended achievement, the direction of travel is clear. Outcomes will of course depend on how much public funding is available, though. A well-integrated, high-quality public transport system requires subsidy. Where the money will come from to level up local public transport is not yet apparent – how much will come from central government and how much from local sources, such as road user charging. The White Paper is positive about devolving powers to local leaders and committing to forward funding streams.

The focus on improving public transport is sensible as a levelling up objective since the high standards achieved in London are visible and applauded. The omission of any commitment to increasing local road capacity is also sensible since there is no general relation between road connectivity and economic well-being or quality of life, so a levelling up objective would be difficult to specify.

One virtue of the Levelling Up White Paper is that it treats improving local transport as just one of a dozen missions that are necessary to tackle regional inequalities, all of which must be pushed forward if the effort is to succeed. Until now, decisions about transport investment have largely been taken in a silo by the Department for Transport, which holds the purse strings and determines how investment decisions should be taken, all while being reluctant to take account of how transport investment can foster development in particular locations since this complicates the modelling. But the intention now is that central government decision-making will need to be fundamentally reoriented to align policies with the levelling up agenda and to hardwire spatial considerations across Whitehall – a welcome development in my view.

All in all, the many recent developments discussed in this chapter offer the prospect of worthwhile incremental improvements to how we travel. Yet the transport system is largely dependent on fossil fuels for propulsion and it therefore makes a major contribution to national carbon emissions, which points to the need to go beyond incremental change. This is the subject of the next chapter.

Chapter 7

How can transport be decarbonised?

The experience of the pandemic has shown that large changes in travel behaviour are possible when they are necessary to protect health, but these will not persist in the longer term to an extent that would make a significant impact on the existential problem of climate change.

The need to reduce and then eliminate greenhouse gas emissions to limit further global warming is well understood. The goal of the Paris Agreement – which was adopted in 2015 and has been signed by more than 190 countries – is to keep the rise in mean global temperature to well below 2 °C above pre-industrial levels, and preferably to limit the increase to 1.5 °C. Each country had to plan and report its contributions. The COP26 meeting in Glasgow in November 2021 kept alive the hope of limiting the global rise in temperature to 1.5 °C, and came up with plans to develop new and updated national emissions targets.

The UK government has committed to reducing emissions by 78% by 2035 compared with 1990 levels, which would put the United Kingdom on a path to net zero emissions by 2050 at the latest, and on a trajectory consistent with the Paris Agreement.

The pre-eminent question for national transport policy is how to decarbonise the transport system. There have been two recent major analyses of the options. The first was from the government's official advisors, the Climate Change Committee

(CCC). It was published in December 2020 and it covers the whole economy, including transport.[1] The second was released by the Department for Transport (DfT) in July 2021 and it deals specifically with decarbonising transport.[2] The government's plans for decarbonising the whole economy, published in October 2021, subsumed the DfT plan.[3]

These exercises have the same ultimate objective of net zero carbon emissions from the whole economy by 2050, with front-end loading so that three-quarters of the reductions are achieved in the first half of the thirty-year period. The starting point is that surface transport is the highest-emitting sector – it was responsible for 22% of total carbon emissions in 2019. The car is dominant, contributing 61% of surface transport emissions and 78% of UK road travel (in vehicle-kilometres). Surface transport emissions have changed little over the past thirty years, so achieving the necessary steep downward trend is a real challenge.

Despite the agreed objective, there are some significant differences in the means to the common end. To reduce transport carbon emissions, the CCC proposes two broad approaches: changing behaviour and developing new transport technologies.

Behavioural change

According to the CCC, behavioural change to achieve a reduction in the demand for car travel operates through three main channels.

- The first is societal, facilitated by technological changes, including increased homeworking and online shopping. This could contribute a reduction in car mileage of between 4% and 12% by 2050.
- The second is increased car occupancy, moving from the current average of 1.6 persons per vehicle to 1.9 by 2050.
- And the third is to shift 9–14% of journeys by car to active travel and 5–8% to public transport by 2050.

The CCC has cited evidence to justify their estimates, but there is still a question mark over them similar to one we have met before: the possibility of optimism bias stemming from taking a view of what is achievable that is influenced by the higher purpose – in this case the need to reduce transport carbon emissions.

Could vehicle occupancy be increased?

Consider, for instance, the evidence that suggests vehicle occupancy might be increased. The CCC suggested that high-occupancy vehicle (HOV) lanes are an example of local interventions that can encourage car sharing. There is extensive experience of HOV lanes on US multilane highways. These lanes allow access only to vehicles with more than one occupant, and consequently they are less congested and journey times are reduced. However, there are few examples of such lanes in Britain, and the ones there are have not all been successful: for instance, a 3.5-mile section of the M4 between central London and Heathrow airport was reserved for buses and licensed taxis between 1999 and 2012 but was then scrapped. It seems unlikely that significant space would be allocated to HOV lanes, given the limited capacity of UK highways. The conversion of the hard shoulder to a running lane – the key innovation of the Smart Motorway concept – would have been an opportunity to create HOV lanes, but no consideration seems to have been given to that possibility.

The CCC also suggested that social pressure may increase car occupancy as the public becomes increasingly environmentally aware, that employers may encourage car-sharing schemes for commuters, and that a variety of shared mobility innovations may play a role, e.g. real-time car-pooling apps.* While all these

---

*Indeed, there are well-established examples of these apps, with examples including Liftshare, which facilitates shared car journeys to work, and BlaBlaCar, which arranges carpool rides to share costs and travel in company.

influences that might increase vehicle occupancy are possible and desirable, the question is the extent to which we can rely on their contributions to cut carbon emissions. Shared commuting car journeys increase occupancy but reduce flexibility of working time, and they are therefore best suited to those who work fixed shifts. There is also a lack of flexibility to divert to another destination – shopping on the homeward commute, for example. The advantages offered by single occupancy, in terms of both timing and destination, constitute a barrier to change that is difficult to overcome.

As discussed in chapter 4, the advent of digital platforms has facilitated car sharing and a variety of other means of reducing individual car ownership and use. Digital platforms allow supply and demand to be met online in virtual markets, and they are having a huge impact in traditional retail markets. For travel, we have seen the transformational impact of ride-hailing, exemplified by Uber, on city taxi services around the world. We have also seen a transformation in the booking of rail and air travel, reducing the role played by travel agents and ticket offices. At one time it looked as though dockless bicycles, operated via smartphone apps, would make a big impact on urban travel, but the attraction faded and investors ended up losing money. As well as shared car ownership and use, other travel modes can be facilitated by apps: e-bikes, e-scooters, demand-responsive minibuses and Mobility-as-a-Service being some examples. Demand for all of these is growing only slowly (at best) though, suggesting that the markets being developed are fairly niche. It is characteristic of new market opportunities facilitated by digital platforms that the possibility of rapid growth allows entrepreneurs to attract substantial funding from investors hoping to profit from first-mover advantage. Conversely, the slow growth that has been seen suggests that market transformation is unlikely, and that these alternatives to the car cannot be counted on to contribute much to transport decarbonisation.

Less car use?

The CCC is optimistic about achieving modal shift from car to active travel (walking and cycling) and to public transport. I discussed this earlier in the context of cycling in Copenhagen, where car use is only slightly lower than it is in London while public transport use is half as high. My conclusion is that we may be able to get people off the buses onto bikes – which are cheaper, healthier, better environmentally, and no slower on congested roads than public transport – but do we want to do that if it means lower income from fares and, therefore, a less frequent service? Note, too, that as buses switch away from diesel engines, the decarbonisation benefit of cycling declines.

What we would really like to do is get people out of their cars, but that is difficult, even in a small, flat city like the Danish capital, where there is excellent cycling infrastructure and everyone has experience of safe cycling. Besides, 80% of carbon emissions from UK car journeys arise from trips of more than five miles, which means that the scope for carbon reduction by switching to cycling is quite limited. Shifting from car to foot is even more difficult, given that walking is the slowest mode of travel and that we are all limited in the amount of time we can spare for daily travel. Walking the shortest trips, rather than taking the car, would be healthier, but it would have little impact on carbon emissions, 95% of which arise from trips of more than two miles.[4]

Modal shift from car to public transport is unappealing for those who own a car, given that the fixed cost of ownership has already been incurred, so the convenience of the car makes it attractive compared with the cost of the bus, in terms of both time and fares. Those with access to a car make little use of buses: people living in a household without a car or van make on average ten times more bus trips than the main driver in a household with a car.[5] The cost of the bus fare is a marginal factor in deciding to use a car, however: when public transport has been made free to use, as has happened in a number of cities around the world, the experience is that of strong passenger growth overwhelmingly drawn from walkers and cyclists, with only marginal effects on car traffic.[6]

Nevertheless, where there is a lack of road capacity – for movement or parking or both – the balance tilts toward public transport. In dense cities with good bus and rail services, car ownership is a cost that can be saved. London therefore has markedly lower car ownership than the rest of England: 44% of London households own no car compared with 20% elsewhere.[7] The Mayor of London's aim is to reduce car use to no more than 20% of journeys in the city by 2041, although this will be difficult to achieve and is beyond the aspirations of other UK cities.

Taken all together, the CCC's plans display quite a lot of wishful thinking about the scope for reducing the demand for car travel. They are based on policy interventions derived from examples of best practice that, it is hoped, would be taken up widely through exhortation and education. The stable situation regarding average travel behaviour that was established in the twenty years before the pandemic suggests that there is a balance between the access benefits of car travel and the time and money costs involved – a balance that it may be difficult to shift.

The government's plan for decarbonisation

The DfT's decarbonisation plan places less emphasis than the CCC's on behavioural change. The report's foreword, by the Secretary of State for Transport, Grant Shapps, is explicit:

> It's not about stopping people doing things: it's about doing the same things differently. We will still fly on holiday, but in more efficient aircraft, using sustainable fuel. We will still drive on improved roads, but increasingly in zero emission cars. We will still have new development, but it won't force us into high-carbon lifestyles.

Behavioural changes are expressed as aspirations:

> We must increase the share of trips taken by public transport, cycling and walking. We want to make these modes the natural first choice for all who can take them.

Quantification of emissions reductions in the DfT's decarbonisation plan is limited, but some headline figures are provided. Increased walking and cycling are projected to reduce greenhouse gas emissions by 1–6 $MtCO_2$ over the 2020–50 period, which is minuscule in comparison with current domestic transport emissions of 122 $MtCO_2$ *per year*. Moreover, the range of outcomes is exceptionally wide, indicating considerable uncertainty about the likely impact of policy and investment. Reductions in emissions from cars and vans are put at 620–850 $MtCO_2$ for the same period, so the savings from active travel would be 1% of that expected from zero-emission vehicles at best. This is remarkable in view of the prominence given to active travel in the DfT's plan, including the intention to invest £2 billion in this area over five years and the aim that half of all journeys in towns and cities will be cycled or walked by 2030. On the other hand, small reductions in emissions from a shift to more active travel would be consistent with the Copenhagen experience, which indicates that increased cycling draws people from buses not cars.

For all the fine words about the desirability of active travel in place of short car trips, then, the projection of carbon savings is inconsequential, perhaps reflecting virtue signalling by policy makers, keen to be on the side of the angels in the face of scepticism from modellers.

Increasing the cost of motoring?

The DfT is very largely relying on technological change to reduce carbon emissions, as I will discuss further below. There is inevitable uncertainty about what may be achieved, and the DfT recognises that additional measures beyond those identified in the present plan may be needed to support the final transition to fully zero emission surface transport. There are hints that such measures may include changes to the cost of motoring, to ensure both that the tax system encourages the uptake of EVs and that 'revenue from motoring taxes keeps pace with this change, to ensure we can continue to fund the first-class public services and infrastructure that people and families across the UK expect'.

Other than this hint, the DfT plans are silent on changes to costs and prices to effect decarbonisation through behavioural change. The CCC is similarly silent. This is not surprising. While the policy of subsidising EVs to promote uptake is not particularly controversial – albeit that it will involve substantial costs if it is effective – increasing charges and taxes to promote decarbonisation is very likely to be a political hot potato. Yet if it turns out that we need to place greater reliance on travel demand reduction to reduce emissions, some stronger interventions may be needed, and of the available measures, a change in relative costs is likely to be most effective. The DfT admits that over the last twenty years the cost of motoring has fallen by 15% in real terms, while the cost of rail fares has gone up by more than 20% and bus and coach fares by more than 40%.[8] To achieve a significant behavioural shift away from the car, some increase in the cost of motoring will surely be needed.

Petrol and diesel fuel at pump prices includes duty that comprises some 60–70% of the total cost, depending on the ups and downs of the wholesale oil markets. In 1993 the Conservative government introduced the Fuel Price Escalator, which raised fuel duty at a higher-than-inflation rate each year, with the aim of reducing both pollution from vehicles and the need for new road construction. However, major protests by hauliers and others about fuel prices in 2000 led to the withdrawal of automatic increases, and fuel duty has been frozen ever since.

Transport costs amount to about 14% of average household expenditure, and fuel costs account for around a quarter of that total. While motorists will tolerate increases in fuel prices arising from fluctuations in the price of oil, increases in duty provoke opposition, often encouraged by the popular press. In France the *'gilets jaune'* mass protests that began in November 2018 were motivated in part by a fuel tax increase that was intended to combat climate change – the protests continued until the coronavirus pandemic put a stop to them. Governments are therefore very cautious about raising taxes on petrol and diesel, even when there is a persuasive environmental argument for doing so. The impact is particularly acute on those low-income motorists for whom alternatives to the car are not feasible for essential journeys.

As electric vehicles increasingly enter the market, raising fuel taxes becomes especially problematic. Better-off motorists and businesses will be able to afford the higher capital cost of new vehicles and will benefit from their lower running costs, whereas lower-income drivers will continue to rely on older cars powered by the internal combustion engine, incurring fuel tax.

More generally, it tends to be politically difficult for governments to tackle an environmental harm by increasing the tax on a product or service that accounts for a significant part of household expenditure. It is usually more effective to do good by stealth, through regulation that reduces the harm by forcing technological change – the deployment of electric vehicles being a prime example. Here, the costs are less visible to users, and they usually fall as experience is gained and production volumes grow. Nevertheless, the loss of revenue from road fuel duty as electric vehicles increase in number will prompt further consideration of the scope for charging drivers for use of the road. Such charging may change behaviour in ways that we will consider next.

Road user charging

Road user charging – also known as road pricing or congestion charging – involves imposing a monetary charge on vehicles while they move. We are used to paying while stationary at kerb-side parking, and also while on the move when using toll roads or in London's Congestion Charge zone. One argument often made for road user charging is that without fuel taxation or a similar charge related to distance travelled, the running costs of EVs will be substantially lower than those of conventional vehicles, which would result in more miles travelled and thus more traffic congestion. More generally, an increase in use as a result of a reduction in cost is known as a 'rebound effect', as when improved home insulation reduces energy requirements and costs, allowing the thermostat to be turned up for greater comfort. One scenario of the DfT's 2018 Road Traffic Forecasts illustrates this expectation.[9] However, the average distance travelled by car has remained stable in recent years, being limited by the time available for travel, the speed of travel and the proportion of households owning cars, none of which has increased in this century. We should therefore not expect the replacement of the internal combustion engine by the electric motor to have much impact on vehicle use. The rebound effect does not apply here.

Another argument for road pricing is to alleviate road traffic congestion, which is the intention of the London congestion charge. Experience has shown that congestion reduction is quite limited at the level of charging typically employed, particularly in a prosperous city like London, where many are able to afford the charge. Charging for road use benefits those who can readily afford to pay by displacing those who are less able to, leading to increasing inequality in use of the road network, which has historically been a relatively egalitarian domain. Nevertheless, congestion relief is, in principle, a possible aim of the road pricing regime, although the magnitude of the charge would need to reflect both the level of congestion and affordability in the locality if congestion is to be effectively ameliorated. Extension of the same charging technology to discourage use of polluting vehicles has been adopted in London's Ultra Low Emission Zone

(ULEZ), the rationale for which will diminish over time as EVs are increasingly used.

Revenue from the London congestion charge and the ULEZ are retained by the city authority, as are revenues from low-emission zones planned elsewhere. These revenues are ring-fenced for expenditure on transport services, as are revenues from kerbside parking charges. A national scheme for road user charging might bring in revenue both for the Exchequer and for local authorities. The local element of the charge could be fixed to reflect local conditions, including congestion and other environmental impacts of traffic, as well as the need for revenues for road maintenance. Local authorities might be permitted to set their share of the road user charge to cover the full cost of local transport provision, obviating the need for grants from the DfT (other than, perhaps, to level up economically lagging regions). The Exchequer element of the charge could depend on the type of road (it could be higher for motorways, which are funded nationally, for example), and it could vary by region to aid levelling up policies.

Introducing a national scheme of road user charging would be a major policy step. There would be attractions in introducing road pricing for EVs alone, with the rationale being that they should pay their fair share of the costs of the road network that conventional vehicles are paying via fuel duty. However, the lower operating costs of EVs are a necessary incentive to purchase the vehicles while capital costs remain higher. As capital costs fall, as is expected, there would be scope for charging users of EVs by introducing a road pricing regime from which conventional vehicles were exempt. A credible policy approach would be to propose that once the capital costs of EVs fall to the level of their conventional equivalents, the tax levied should be similar for both types of vehicle.

One way of selectively charging EVs would be to introduce a general road pricing system but to credit conventional vehicles with the fuel duty they pay. This is the basis of a voluntary pilot scheme in the US state of Oregon. Alternatively, conventional vehicles might be exempt from the user charge, although this would limit use of the charging mechanism to manage congestion.

Given the political and practical difficulties of overnight national implementation of the necessary payment and enforcement technologies for road user charging, it is worth considering options for incremental roll-out. The existing congestion charging system in London has been in operation for nearly two decades now, and it functions sufficiently well, generates useful amounts of net revenue and is publicly acceptable. Its scope is limited, though, by the fixed daily fee for entering the charging zone. There are a number of incremental developments of the London scheme that might be feasible and that would move us towards a scheme of national applicability.

The key innovation would be to charge vehicles via a smartphone app, with the incentive for adopting the app being a discount from the standard daily fee. Smartphones know their location in time and space, and it would be possible to link them to a specific vehicle so that a charge would be levied only when a vehicle that is liable to be charged is in a charging zone at a time when charging is in operation. Charging could then be made more flexible than the fixed daily fee: for instance, the charge could be varied according to duration and location in the charging zone, the time of day or the level of congestion. This should be publicly acceptable so long as the charge made was less than the default standard daily fee. Camera enforcement via vehicle number plate recognition would continue as at present.

Once this smartphone charging system was working reliably, there would be options to extend it to other parts of London where traffic congestion is a problem. The charging and enforcement systems could be made available to other cities that wished to manage local traffic, incentivised by revenues that could be used to provide alternatives modes of travel to the car. And once a number of cities were using road pricing by means of a smartphone app, the basis for national adoption that could be applied to EVs would exist. This would need to be supported by the national roll-out of camera enforcement, unless a better enforcement system could be devised.

Whichever way it was brought about, a decision to adopt national road pricing would need to be presented as part of the strategic vision to phase out internal combustion engine vehicles, which commands wide public support. It would be important to send an early signal that EVs would not be exempt from taxation in the long run. Part of the revenue from road user charging might be employed to fund a scrappage scheme for older internal combustion engine vehicles, to speed up decarbonisation of road transport.

More generally, there is support for tackling climate change across the political spectrum in Britain – unlike in the United States, where the issue is very divisive. Wide support needs to be maintained if the Net Zero objective is to be achieved. An opinion survey carried out in 2021 found that two-thirds of UK adults believed that climate change is either a dangerous or an immediate threat, but that solutions must be fair, involving a determined-but-steady transition.[10] Any perceived unfairness in bearing the costs of the transition would offer opportunities to populist politicians to resist change, which could undermine success in achieving the objective. A 'just transition' is needed.

The countries of the European Union are considering proposals from the European Commission to become 'climate-neutral' by 2050, achieving a 55% reduction in emissions by 2030 compared with 1990 levels. Measures include creating an emissions trading scheme for fuel supply for buildings and road transport. An emissions trading scheme puts a cap on a sector's emissions, with the cap reducing over time; businesses must acquire emissions allowances to match their emissions, and they can buy or sell these at a market price. An emissions trading scheme for road transport would put a price on carbon emissions from these sources and would therefore add to the cost to users.[11] This would be contentious, but were the European Union to succeed in imposing such an additional cost on the use of carbon-based fuels for road transport, it would suggest that this might also be possible in Britain.

Technologies for decarbonisation

Changing travel behaviour is one approach to reducing transport greenhouse gas emissions. The second kind of approach to decarbonising the transport system is to rely on technological innovation, and to achieve the decarbonisation, through government intervention, at a faster rate than would occur from purely commercial decisions. Both the CCC and the DfT have identified a range of developments that are necessary to track the route to Net Zero.[12]

How fast will electric vehicles roll out?

By far the most important technology for decarbonising transport is electric propulsion for cars and vans (we looked at this technology in chapter 4). Here I focus on the factors that influence the rate of market penetration.

All car manufacturers are developing and marketing EVs, and sales are increasing. While the pioneering Tesla models have been relatively expensive, the established auto makers are now increasing volumes and driving down costs. The basic version of Volkswagen's all-electric ID.3, similar in size to the company's Golf, is claimed to be less expensive to buy than comparable internal combustion models.

Sales of EVs are inhibited by perceptions of limitations to the distance that it is possible to travel between recharging and the availability of charging points ('range anxiety'). These constraints will lessen as battery performance improves and as investment in charging infrastructure progresses. Access to better batteries is a crucial source of competitive advantage to the car manufacturers, so the incentive to advance the technology is strong.

In the long term, EV charging away from home will be the responsibility of private sector businesses, as for other sources of energy. But in the near term, demand for charging may not be large enough to justify sufficient private sector investment to counter range anxiety. A chicken-and-egg situation therefore arises whereby demand growth is inhibited by the perceived lack of charging facilities. Accordingly, there is a role for governments to promote the development of the charging infrastructure, to give confidence that longer trips can be made without mishap. Analysis by the Competition & Markets Authority concluded that by 2030 there will need to be at least ten times more public charging points than there currently are (around 25,000), both on-street kerbside charging for those without a driveway and rapid charging for longer journeys. There are also concerns that need to be addressed about the reliability of public charge points, and the excessive variety of means of payment.[13]

The CCC's estimates of the attractiveness of EVs are based on credible assumptions about the factors mentioned above and the likely costs involved, and it has concluded that battery EVs will have reached upfront cost parity with internal combustion vehicles by 2030.* With the fuel cost savings from electric propulsion, battery EVs should be sufficiently attractive to amount to more than 90% of new car and van sales by 2030, which is consistent with the government's declared intention to prohibit the sale of internal combustion engine cars and vans by that date.

Most private purchases of new cars involve financing arrangements with monthly payments. This allows car buyers to understand the trade-off between higher purchase costs for EVs and lower operating costs, which may be helpful in incentivising the switch from conventional vehicles. The contribution of EVs to transport decarbonisation depends in part on how long

---

*The limitations of plug-in hybrid EVs were discussed in chapter 4.

people hang on to ageing petrol and diesel cars. This will in turn depend on the continued availability of fuel supplies as demand declines, and also on the availability of servicing, given that EVs require less effort to maintain than the much more complex and stressed internal combustion engines. As noted above, a scrappage scheme to retire the most carbon emitting conventional vehicles might be funded from a road user charge, to accelerate their phase-out.

Other routes to vehicle decarbonisation

Beyond electrification of cars and vans, there are emerging opportunities for the electrification of heavy vehicles. Electric buses are already in operation in a number of UK cities, with overnight recharging at depots. For heavy goods vehicles (HGVs), the weight of the battery necessary for normal distances detracts from the load that can be carried. An alternative technology is the hydrogen fuel cell, which would allow refuelling with hydrogen either in-depot or from filling stations. This would require development of infrastructure for hydrogen manufacture and distribution. The route to decarbonisation of HGVs will involve competition between battery electric vehicles and hydrogen fuel cell ones, and the outcome is unclear at present.

Biofuels from sustainable resources can be added to, or can replace, conventional oil-based fuels. The petrol available on British forecourts has long included up to 5% bioethanol, and this has recently been increased to 10%. Biodiesel, largely derived from used cooking oil, is also increasingly incorporated into road fuel, and this might be particularly useful for decarbonising HGVs.

Only about 40% of the UK's rail network is currently electrified, with the remainder being used by diesel trains. There is scope for improving the efficiency of diesel propulsion and to electrify up to 60% of the rail network. However, only a small proportion of rail freight is hauled by electric locomotives due to the need to travel on routes that are not widely used, where electrification would be uneconomic. Battery electric and hydrogen propulsion technologies are being developed for this purpose.

In the long run, surface transport decarbonisation is feasible, with electric or hydrogen propulsion replacing oil-based fuels. It will be necessary for the electricity supply system to be decarbonised, relying on renewables and nuclear to replace gas, and it will also be necessary for the materials used in surface transport to be made in ways that do not emit carbon dioxide to the atmosphere (steel and cement manufacture are the biggest culprits here). All these requirements are implicit in the commitment to achieve Net Zero by 2050.

Decarbonising air travel

Air travel is much more of a problem. Aviation has experienced substantial growth in passenger demand, and the CCC has considered scenarios for future growth to 2050 in the range from −15% to +64%, with +25% being assumed for the achievement of Net Zero overall.

Demand management for air travel is best achieved by preventing airport expansion, which in any event tends to be locally unpopular on account of aircraft noise, with the proposal for a third runway at Heathrow being the most prominent and controversial example. The general case made for airport expansion is to support economic growth – by making more destinations directly accessible to UK exporters and inward investors – and to enhance the status of London as a world city for doing business. However, as I discussed earlier, business travellers comprise only a minority of air passengers – even at Heathrow the figure is only some 25%. There is therefore ample space for business travel to expand if demand grows by displacing leisure travellers to other airports with spare capacity.

While airport capacity constraints would eventually limit passenger demand *growth*, the scope for demand *reduction* is generally regarded as quite limited. The CCC recommended that, first, there should be no net expansion of UK airport capacity unless the sector were on track to sufficiently outperform its net emissions trajectory and could accommodate the additional demand; and, second, that a demand management framework should be developed to annually assess – and, if required, control – aviation emissions. In responding to the CCC, the DfT asserted that flying is a social and economic good that the government wholeheartedly supports, and it said it believed that the aviation sector could achieve Net Zero without the need to intervene to limit growth. The DfT further asserted that there are scenarios in which the Net Zero target could be achieved by focusing on new fuels and technologies rather than by capping demand.[14] This is a decidedly optimistic view, and it is not supported by any analysis of technology options commissioned by the DfT.[15]

Given the difficulty of restraining demand growth, there is great interest in new aviation technologies, including the possibility of electric propulsion – taking advantage of progress in battery technology for road vehicles – which would permit safer, quieter, multiple-rotor vertical lift aircraft for use in urban airspace. There are many models of the latter in experimental development, and they might be able to replace conventional aircraft for short trips. Hydrogen is another potential energy source for short-distance air travel. For long-distance air travel, biofuels from a variety of sources, also known as sustainable aviation fuel (SAF), could be added to, or eventually replace, jet fuel.[16] Other possible SAF sources are electrochemical fuels formed by combining hydrogen with CO_2 to form a gas which is then converted into a liquid fuel.[17] These new propulsion technologies will take time and effort to develop, and the ultimate operational costs for aircraft are likely to be higher – possibly substantially higher – than for present operations using kerosene.

Carbon emissions from aviation amount to about 3% of global CO_2 emissions from fossil fuels at present, but the level was growing quite fast before the pandemic struck.[18] Replacing oil-based fuel for air travel is much more challenging than it is for surface travel. Any remaining carbon emissions from aviation by 2050 would require the development of technology for the removal of greenhouse gases from the atmosphere. Indeed, aviation is likely to be a key driving force behind the long-term deployment of engineered carbon removal.

The aviation industry has been late to engage seriously with its contribution to climate change. Efforts are now underway to develop alternatives to oil-based fuels. One future policy option would be to force development and uptake of suitable technologies by regulation once feasibility is established, as is being done for zero-carbon road vehicles. This would allow higher costs to be accommodated on a level playing field for airline operators. However, it is too early to assess the likely timing of any of the alternatives becoming feasible. Nevertheless, there are precedents for rapid development, cost reduction and wide deployment of new technologies when there is both a need for them and sufficient resources are committed. Current examples include the coronavirus vaccines, photovoltaics, wind power and electric vehicles, with Elon Musk exemplifying the opportunity for an entrepreneur to lead innovation to respond to climate change. Longer ago there was the speedy development of the atomic bomb, the jet engine and the V2 rocket during World War II, and subsequently of nuclear power.

Although the reluctance of the government to contemplate policies to limit the growth of air travel is understandable – and all the more so if the target were to be to reduce future demand – such an approach may yet be needed if the development of technologies to decarbonise aviation is too slow.

Predicting the future

While many promising technological developments are underway, the question is to what extent these advances in technology can be relied upon to deliver transport Net Zero by 2050. Projections of the impact of both technological developments and behavioural change are based on models whose validity is unproven. As I discussed earlier, small-scale traffic models can be used to predict short-term traffic flows and inform interventions such as changing the timing of signals. The observed responses to such interventions can be used to refine and validate the model. However, that process gets more difficult as models are elaborated: scaling up spatially to cover regions or nations; including all modes of travel; looking forward sixty years to estimate the benefits of new investment; and incorporating projections of the emissions-related behavioural and technological changes discussed above.

The resulting models are complex and opaque, with many parameters whose values require the exercise of judgement. This means the models are open to bias, and particularly optimism bias, which arises when modellers make choices, consciously or unconsciously, that tend towards achieving a strategic purpose. Modellers naturally want to please their clients and may, within the bounds of professional respectability, select parameter values to generate outcomes that meet client expectations. Such expectations may also be coloured by biases – to favour technological solutions, for example – or, conversely, be affected by scepticism about the likely development of new technologies.

Bias in modelling

As I mentioned in chapter 3, optimism bias is a well-recognised phenomenon in transport modelling. The CCC methodology for modelling future transport carbon emissions assumes that we achieve the 'highest possible ambition' for reducing emissions while taking into account real-world constraints such as the need to build up skills, supply chains and manufacturing capability, and the economic lifetime of existing, high-carbon assets.

This seems like a recognition of optimism bias, motivated by the formidable challenge of reaching Net Zero by 2050.

One way of countering bias that has been adopted by the CCC is to explore a range of scenarios that reflect possible extents of public engagement and technological innovation. These apply to the whole economy and include scenarios that reflect optimism and pessimism about both behaviour change and technological innovation. However, multiple scenarios are unhelpful for those charged with making decisions. Accordingly, the CCC narrowed down its focus to a single 'balanced' pathway to Net Zero as the basis for its recommendations, thereby resurrecting the question of bias.[19]

Unlike the CCC, the DfT provides little in the way of detailed projections of the impact of the different policy and technology elements that are expected to achieve transport sector Net Zero. Rather, a range of carbon reductions is shown over time, as indicated in figure 7.1 for domestic transport. The spread of pathways is stated to reflect uncertainty over policy design, economic growth, fuel prices, travel behaviour and volatility in emissions, with the range narrowing over time due to the increasing proportion of zero-emission vehicles.

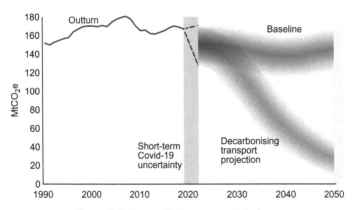

Figure 7.1. Domestic transport emissions.
(*Source*: DfT, Decarbonising Transport, 2021.)

The DfT's decarbonisation plan takes a more optimistic view of the potential of technological developments than does the CCC. Although prominence is given in the narrative to the promotion of active travel, the assumed carbon reductions from changes to travel behaviour seem quite low, with reliance placed on new technologies to get to Net Zero. Such a viewpoint is not unique. An analysis by the independent International Council on Clean Transportation found that technological progress could very largely achieve reductions in global transport carbon emissions to limit warming to 1.5 °C.[20] On the other hand, modelling conducted under the auspices of the UK Energy Research Centre, a consortium of university researchers, suggests that there is a need for much greater policy emphasis on travel demand reduction and the changes to lifestyles that would be required.[21]

The DfT might be said to be taking the path of least political resistance: taking an optimistic view of the potential of decarbonisation through technology and playing down the potential need for significant changes in travel behaviour. The virtue of this approach is that it makes possible a broad political consensus. Moreover, such a tactic is not unreasonable, in that uncertainty about the validity of modelling means there is no secure basis on which to initiate potentially unpopular behavioural change. However, technological developments commonly take longer than initially hoped, and the question accordingly arises of whether the decarbonisation of surface transport is happening at a fast enough pace to achieve the early reductions necessary to avoid warming above 1.5 °C, as required by the Paris Agreement. The DfT has therefore committed to producing a further transport decarbonisation plan within five years, to ensure transport is on the correct pathway to achieve Net Zero.

In the meantime, it would be desirable to get more agreement among the modellers about outcomes, and particularly about the importance of changes in travel behaviour to get to Net Zero. Transport modelling is complex and opaque, so it is hard to understand the significance of the whole variety of assumptions that must be made about future economic and behavioural relationships. Much of such modelling is carried out by consultants using proprietary software. The DfT's National Transport Model, the basis for its projections of carbon emissions, employs proprietary software, and this is the reason given for not making its model public, meaning that others cannot use it with their own assumptions (note, though, that a new version of the model is under development that would be open to other users). This lack of transparency is commonplace for transport modelling, and it is in contrast to modelling exercises in other areas. For instance, climate change modelling is an international, collaborative, transparent effort that is the main input to the work of the Intergovernmental Panel on Climate Change. Nearer to home we have the epidemiological modelling of the coronavirus pandemic that has informed government policy on lockdowns, vaccination strategies and the rest, carried out not within government but by a number of university groups in an open and collaborative manner.

We need major reform of the approach to modelling transport carbon emissions to achieve a more coherent view of the outcomes of measures both to deploy new technology and to change travel behaviour. This is important because more strenuous measures to change behaviour, beyond the encouragement of active travel, may meet resistance, either because they reduce the amount of access that the car and air travel has made possible or because they increase the cost of travel.

Transport modelling therefore needs to move on and become transparent and collaborative rather than opaque and proprietary. More effort needs to be devoted to validating models by comparing forecast with outturn where that is possible, e.g. over the initial years following the opening of a new element of infrastructure. For the period through to 2050, the best that can be done is for modellers to run their models on common assumptions, to understand why forecasts differ, and then to vary assumptions to test the sensitivity of forecasts to bias (both optimism bias and pessimism bias), whether it concerns technological innovations or behavioural change.

Reaching decisions on decarbonisation

At present there is a broad consensus about the policies needed to tackle climate change in Britain, but there is no guarantee that this would be sustained if additional measures, more disruptive to current lifestyles, were proposed. Those of a populist disposition, both politicians and press, would likely resist, playing down the urgency of action; hoping that breakthroughs in technology would come to the rescue, including carbon capture and storage and geoengineering; and pointing up slow progress by other major emitting countries that are still building coal-fired power stations. Moreover, while the practicalities and costs of decarbonising surface transport are becoming understood and seem reasonably affordable, this is not yet the case for decarbonising home heating (by replacing gas with heat pumps and/or hydrogen). Nor is it the case for air travel, which is the biggest problem for the transport sector.

One innovative approach to reaching decisions that has been tried in recent years has been to convene a 'citizens' assembly': a representative group of citizens who are selected at random from the population to learn about, deliberate on and make recommendations in relation to a particular issue or set of issues. It is still up to elected politicians whether or not to follow an assembly's recommendations. Typically, up to a hundred volunteers meet together over two or more weekends, with expert advisors available to provide input and facilitators on hand to help structure the discussions.

In 2020, six House of Commons select committees convened an assembly to understand public preferences on how to tackle climate change. The assembly's headline recommendations included a future that minimises restrictions on travel and life-styles, placing the emphasis on shifting to electric vehicles and improving public transport rather than on large reductions in car use.[22] This aligns broadly with the DfT's subsequently published decarbonisation plans. However, in the view of one of its expert advisors, the assembly's stated conclusion did not adequately reflect members' willingness to contemplate more radical changes to achieve zero carbon emissions from surface transport by 2050.[23]

I myself was involved as an expert advisor to a citizens' assembly held in Cambridge in 2019 to address the problem of road traffic in the city. I was impressed by the commitment of the members and the quality of the discussion and debate. The recommendations included restricting car use in the city centre, in part by means of a charging scheme that would provide finance to support better bus services.[24] Levying a new charge on motorists is, however, a sensitive proposition, and implementation is the subject of ongoing political debate.[25] This points to a general problem with the citizens' assembly approach, which is that a representative group of citizens can come to sensible decisions – given their commitment of time, energy and the availability of expert advice – but that does not mean that other less-well-informed citizens would be bound to agree. Politicians are therefore nervous about simply implementing the sensible recommendations that these assemblies often provide.

Prospects for a more sustainable transport system

There has been a widespread feeling that we should aim for a better world after the pandemic, not just a business-as-usual recovery from the health and economic consequences. The relevant maxim is 'never let a good crisis go to waste', yet in practice the huge effort required to manage the pandemic has tended to distract attention from post-pandemic possibilities. Nevertheless, there have been some potentially helpful developments.

The financial crisis faced by bus and rail operators, which required the government to step in with substantial support, created the opportunity for organisational reform that may allow a more integrated, sustainable public transport system to be developed. Many local authorities took a lead in boosting active travel, mostly by bringing forward schemes that were already in the pipeline, but they often found that it was not easy to carry their citizens with them. The government has given a strong steer, and some cash, to local authorities to support investment in active travel, yet there is limited scope for getting people out of their cars to travel under their own steam, as the experience of Copenhagen shows.

The main constraint on changing travel behaviour is the level of access to people and places to which we have become accustomed. Access is limited by the time we can spare to be on the move, and it depends on being able to travel over the required distances at speeds that are feasible by car or good public transport.

We have inherited the towns and cities built by our predecessors, which in general we seek to preserve and, in any event, could not afford to rebuild. Accordingly, access is also constrained by the existing built environment, within which sit our journeys' origins and destinations, and by the existing location of transport infrastructure. This limits the scope for remodelling cities to create '15-minute neighbourhoods' in which residents can have all their needs met within a walkable distance of their own front doors.

The car is very popular because, on uncongested roads and when parking is available at both ends of the journey, it provides speedy and convenient door-to-door travel. But in urban environments these conditions may not be met, and the consequences for the population at large are detrimental. Other travel options then become attractive, particularly rail and the active travel modes of walking and cycling. Beyond city centres, though, the car will surely remain attractive, but increasingly it will be electrically propelled.

The question prompted by the pandemic is to what extent our journeys are really necessary. The answer is that we were able to manage with substantially less travel, the reduction reflecting our desire to minimise exposure to a very infectious virus. Yet as the pandemic recedes, it seems likely that our desire to travel will largely revive, in terms of both daily journeys from home and longer leisure trips, albeit with some changes in the patterns of movements. There may be some helpful reduction in commuting and in business travel, although to what extent in the long term is unclear, as is the likelihood of other travel arising to replace the time thereby saved, given the long-run experience of unchanged average travel time of an hour a day. If this replacement travel is by active modes, that will contribute to transport decarbonisation, whereas it would be unhelpful if less flying for business meant more flying for leisure trips.

Changes in travel behaviour to reduce carbon emissions will not come about easily, although the pandemic has shown that big changes are not impossible. The alternative is to look largely to advances in technology to meet the climate change objectives for the transport sector. This dichotomy reflects the wider debate that contrasts 'techno-optimism' – the belief that technological progress will enable us to reach zero carbon emissions while continuing to enjoy our existing standard of living – with the 'end of consumerism', which insists that current rich-country living standards are unsustainable in respect of demands for energy and materials.

Adair Turner, the first chair of the CCC, argues that in the long run there is no limit to the amount of low-cost green electricity we can sustainably produce from renewable sources, but in contrast there are severe and immediate constraints in other sectors, particularly food and textile production that depend on inefficient photosynthesis.[26] For transport, this means that we will be able to electrify surface transport at operational costs at least as low as those for fossil fuel systems, and we should be able to decarbonise aviation, shipping, steel and cement by

2050 using renewable or nuclear electricity and hydrogen. However, Turner argues that we have left it dangerously late to move
away from fossil fuels, given present atmospheric levels of CO_2,
and we therefore need to reduce emissions by around 50% in the
next decade if we are to avoid catastrophic global warming. Yet
it would be difficult to increase the rate of introduction of new
technologies to decarbonise transport over this timescale. In
addition, as I have argued, the prospects for achieving useful
carbon savings from changes to travel behaviour are not promising. At the same time, other sectors of the economy, particularly
home heating and food/agriculture, face more substantial problems than transport does in reducing greenhouse gas emissions.

It therefore seems as though we have left it too late to
achieve a pain-free path to Net Zero for transport. There is
much impetus behind technological developments, but uncertainty remains about delivery timescales. Debate will therefore
continue about the need for stronger measures to achieve a
useful reduction in travel demand, e.g. by means of a significant
increase in the cost of travel. Related to this is the question of
what measures would be politically feasible for a democratically
elected government functioning in a global economy in which
there is great diversity of decarbonisation achievement.

The story I have been telling about why and how we travel
is complicated, involving human behaviour, economics, demog-
raphy, technology, public policy, looking back to the past and
projecting forward to the future. In the book's next, and final,
chapter I summarise what we know about how to improve the
transport system, but there remains much we do not yet know
about how to decarbonise our means of travel.

Chapter 8

Is your journey really necessary?

Travel and transport are part of our daily lives, and we naturally have opinions about how things could be improved based on our own experiences. Yet, as I have set out in this book, the system is complex, and simple ideas for improvement may have unintended consequences. Advocates of new investment in roads, railways and airports are optimistic about the benefits, for both users and businesses, while critics may be better able to anticipate adverse consequences. Wishful thinking is commonplace, particularly about investment costs that often significantly overrun initial estimates.

To tackle the problems of the transport system we need to ask two questions, in the following order. First, what's going on here? And second, what can we do about it? We need analysis before advocacy if we are to improve things. The two broad areas offering potential for improvement are adopting better transport technologies and changing our travel behaviour. We are interested in changes in behaviour that reduce the impact of our travel on the environment and on fellow travellers, hence the question in the chapter title above: is your journey really necessary?

My intention in this chapter is to summarise succinctly the key ideas discussed in the earlier chapters, with the aim of helping to analyse the problems we face in order to achieve solutions that work. The approach is intentionally simple, as opposed to rigorously theoretical. There is a vast amount of mathematical analysis of transport economics and modelling out there, but it generally involves simplifying assumptions that have disappointing consequences for real-world outcomes, missing the wood for the trees. What I offer are heuristics based on experience – mental short cuts; rules of thumb – that are intended to be helpful for practical decision takers.

Trends over time

♦ The history of transport over the past two centuries is characterised by the development of technologies that exploit the energy of fossil fuels in order to travel faster: first coal to power the steam engines of the railways; and then oil in the internal combustion engine on the roads and the jet engine in the air. All of these major innovations allowed step-change increases in the speed of travel.

♦ The average speed of travel ceased to increase at the end of the last century. Since then (and prior to the pandemic) we in Britain have travelled around 6,500 miles per person per year by all surface modes, spending an average of about an hour of our time each day travelling, and making about a thousand journeys per year.

♦ We travel to gain access to places beyond our homes. How far we travel is limited by time as well as speed. Average travel time has held steady at close to an hour a day for at least fifty years (the period for which we have data), and probably since humans ceased to be hunter-gatherers and settled in farming communities and towns.

Access and choice

♦ Increased access through faster travel allows us to have more opportunities and choices – of jobs, shops, schools and the rest. Access and choice increase with the square of the speed of travel since the locations we can reach in a given time are defined by the area of a circle whose radius is proportional to speed. Switching from walking at 3 miles per hour to cycling at 10 miles per hour therefore increases access and choice by some tenfold, while a car travelling at 20 miles per hour provides forty times more opportunities than walking. This is the main reason for the popularity of the car.

♦ Access and choice are subject to diminishing returns. The more choice you have of some particular service – supermarkets, say – the less the value of each additional store. The combination of access increasing with the square of the speed of travel but also being subject to diminishing returns therefore means that travel demand can be expected to cease to increase: that is, demand can saturate.

Impact of time constraints

♦ Invariant average travel time of an hour a day is unlikely to change in the future (pandemic shocks excepted). This time constraint is important for how travel behaviour changes as the result of changed circumstances.

♦ The time constraint limits the growth of traffic congestion. Congestion arises in places where there is high population density and high levels of car ownership, such that there are more trips that might be made by car than there is road capacity available. If traffic volumes increase, delays also increase, and some potential road users make alternative choices: a different route or time; a different mode of travel where there are options; a different destination (shopping somewhere else, for instance); or not to travel at all (shopping online, for example). A congested equilibrium is established, with gridlock arising only under unexpected conditions such as roadworks or a motorway pile-up. The scale of congestion depends on what alternatives are available, with urban rail important for limiting traffic congestion in cities.

♦ The existence of a congested equilibrium in which some car trips are suppressed by the prospect of unacceptable delays is the reason why measures to mitigate congestion fail. We know from experience that we cannot build our way out of congestion by adding road capacity, because this simply permits previously suppressed journeys to be undertaken. More capacity allows more trips and traffic, but congestion is not lessened except for immediately after the opening of new capacity, before the new congested equilibrium is established.

♦ Conversely, subtracting road capacity, as happens when cities make more space for buses and bicycles, reduces the volume of traffic without increasing congestion beyond the initial period when road users are adapting to the new delays. It is therefore difficult to change the intensity of congestion on roads open to general traffic, but it is possible to change the amount of congestion in a given area, either by adding or reducing road capacity, which adds to or reduces the volume of congested traffic.

♦ Other measures to mitigate congestion cannot be expected to succeed when there are suppressed car trips. Measures that are often advocated include investing in public transport use and active travel, consolidating freight deliveries into fewer vehicles, and congestion charging (as is already operated in London). Sufficiently high levels of charge for road use could deter sufficient numbers of road users to relieve congestion, but this would likely be problematic for reasons of equity and public acceptability.

Car ownership and use

♦ Road traffic congestion limits the speed of travel by car. The average distance travelled by car has not increased over the past twenty years. This reflects our general inability to reduce congestion, as well as speed limits set for reasons of safety. More generally, we have attained a steady state on a per capita basis since the turn of the century, with average distance, average travel time and average number of trips per year having ceased to grow. This contrasts with the growth in average distance travelled in the last century, which was a result of the growth of car ownership and the addition of road capacity.

♦ Household car ownership in Britain stopped growing twenty years ago, by which time three-quarters of households owned one or more cars. There has been some continued growth in car ownership *within* car-owning households, but the first car makes the main contribution to car use.

♦ Car ownership and use are lower in densely populated cities, where congestion and parking constraints limit use of a car. High population density in cities arises from the concentration of businesses that benefit from agglomeration, co-location that promotes learning, sharing and matching. High population density means that public transport is economically more viable and catchment areas are smaller, whether of schools or supermarkets, allowing more access by walking and cycling. High population density also means lower per capita carbon emissions, both because of lower car use and because homes are smaller and have lower energy requirements.

♦ Successful cities with growing populations cannot increase road capacity without damaging the attractions of their city centres. This leads to the share of journeys by car declining over time – a process facilitated by investment in public transport, particularly rail, which is fast and reliable. Protected routes for cycling increase the number of people using bicycles, but generally by attracting passengers off buses rather than out of cars.

♦ Beyond cities, the car is likely to remain the dominant mode of travel, despite concerns about its undesirable impacts. Cars' tailpipe emissions, which contribute to climate change and urban air pollution, will be remedied as the internal combustion engine is replaced by electric propulsion. Casualties from crashes can be reduced through technological advances and policy interventions. There is some scope for managing, but not eliminating, both congestion and the conflict for space between cars and people. Overall, the attractions of door-to-door travel by car mean that we will put up with these detriments.

♦ While per capita travel has held steady since the turn of the century (prior to the pandemic), total travel demand increases as population grows. If growth is accommodated in new housing on greenfield sites, car ownership and use increase, but if growth takes place in urban environments, the population density of which increases, the resulting travel demand can be met by public transport and active travel. The growth in total travel demand therefore depends on planning policies, both national and local.

Impact of new technologies

♦ New technologies are changing the car and how we use it. Electric propulsion is beginning to replace the internal combustion engine, motivated by the need to eliminate tailpipe emissions that contribute to urban air pollution and to global warming. A change in propulsion technology will not make much difference to how cars are used. The average distance travelled is constrained by time, speed and household ownership, none of which would be increased by electric propulsion.

♦ Digital platforms are changing aspects of travel for the better, allowing convenient online booking of travel by rail and by air, as well as the ability to summon a taxi via a smartphone app. There are a number of other applications that could reduce the attractions of personal car ownership, including shared car ownership and use, and various forms of shared micromobilty that could make public transport more attractive by providing for the 'last mile' to home. The most complex of such app-based offerings is Mobility-as-a-Service (MaaS) that offers a wide range of travel modes. As yet, the commercial viability of these innovations is unproven, as is their impact on personal car ownership.

♦ Digital navigation employing satnav technology allows us to find our way when we travel and offers alternative routes and modes of travel. On the roads, routing can take account of traffic congestion. This can help us use the road network more efficiently but it can also send vehicles into places where through traffic is not desired. A useful benefit of the technology is the prediction of journey times, both in advance of a trip and while en route. This mitigates the main problem arising from traffic congestion: the uncertainty of when you will arrive.

♦ Vehicle automation is seen by many as a natural development of the car, with benefits from improved safety, increased road capacity and less driver effort. Yet the difficulties of general deployment of driverless vehicles are substantial, particularly when it comes to demonstrating safe performance. It is also far from clear that the benefits to users will justify the additional costs.

♦ None of these new major transport innovations – electric propulsion, digital platforms, digital navigation, automated vehicles – offer step-change increases in the speed of travel, unlike earlier innovations such as the steam train, the bicycle, cars and two-wheelers powered by the internal combustion engine, and the jet airliner. The new innovations therefore improve the quality of the journey and/or of the environment without increasing the distance travelled or access to destinations.

Vehicles and people

♦ Urban streets are both places for the movement of vehicles and places where people engage with one another. There are options to increase engagement by restricting traffic, e.g. by increasing pavement space or creating Low Traffic Neighbour-hoods, but these depend on gaining the assent of the residents of a locality.

♦ The most effective measure to reduce car use in the centres of towns and cities is to limit parking provision, through both higher charges and prohibition of on-street parking.

♦ Cycling is good for people's health and for the environment, but boosting cycling draws people from public transport not out of their cars.

♦ The car is an attractive mode of travel where there is road space to move without experiencing too much congestion and where parking is available at both ends of the journey. Given the existing built environment within which journeys are made, and given the access benefits experienced from car-based travel, it would be difficult to achieve much mode shift away from the car.

♦ In densely populated urban areas, congestion and parking availability limits car use. Rail is then attractive because of its speed and reliability, if it is well managed using modern tech-nology. Buses can also be attractive if they are kept segregated from general traffic. The attractiveness of public transport is fur-ther increased if there is integration of buses and trains, offering frequent services with well-understood timetables, interchange hubs and cashless payment systems.

Transport investment

♦ The conventional approach to decisions about public invest-ment to add capacity to the road system involves estimating the economic benefits that are supposed to largely reflect the saving of road users' travel time. However, such time savings are only short term. Adding additional capacity attracts new traffic, which restores congestion to what it had been and so negates the time savings. The extra traffic may mean that people are making more or longer journeys, which generate access bene-fits, yet because access is subject to saturation such additional access benefits are limited. The extra traffic may also comprise local users rerouting to save a few minutes on short journeys, displacing long-distance business users for whom increases in the capacity of major roads are intended.

♦ Proposals for transport investment that rely on notional sav-ing of travel time as the main economic benefit should therefore be viewed with scepticism. So should any models for which the outputs are time savings. Appraisal of proposed investments needs to address the observable changes in travel behaviour that contribute to the main policy objectives: boosting eco-nomic growth, accommodating changing demographics, and mitigating environmental and health harms.

♦ Given the speed and reliability of travel on the railways, unim-peded by road traffic congestion, there is a case for investment in the rail system to modernise and expand services, and to level up the quality of service across the country. The railways can generally cover operating costs from fare revenue but not the capital costs of new construction, which must therefore be funded by government.

♦ There is a good case for rail investment where restricted capacity limits economic development, the best examples of which have been in London. The argument that new rail investment will stimulate economic growth in lagging regions – despite being one that is often made – is less convincing. Where public funding is limited, as is usually the case, improvements to rail services within cities that boost agglomeration benefits are to be preferred to improvements in services between cities.

♦ For any kind of transport infrastructure investment, attention must be paid to *which* users will benefit. Decisions about timetables and fares allow substantial influence over use of the railway. Roads are open to all, and hence local users are able to take advantage of increases in capacity intended for longer-distance business users. Investment in airport capacity permits growth of leisure travel even though the case for investment is concerned with the benefits to business users and the economy.

Are all our journeys necessary?

♦ The coronavirus pandemic showed that we could manage with less travel and less access, motivated by concerns for our health. Once we feel the virus is under control, it is to be expected that we will again seek the access to which we have become accustomed, although there may be changes in travel patterns to reflect changed behaviours such as more working from home and more online retail. Working and shopping from home allow more time travel for other purposes, which could be active travel in the locality, with lower carbon emissions.

♦ Neighbourhoods that offer a good range of services reduce the need to travel further afield and lessen the disadvantage of those who do not have access to a car. A positive approach by the planners is helpful in encouraging mixed-use neighbourhoods.

Changing behaviour

♦ Altering relative prices is the main way in which travel behaviour may be changed, whether by using taxes or subsidies. Car ownership and use would be reduced by increasing the duty on petrol and diesel fuels. Higher fuel costs increase the attractions of smaller, lighter vehicles.

♦ Low operating costs are an attraction of electric vehicles, compensating for their present higher purchase costs. As purchase costs reduce over time, there will be a good case for introducing a road user charge for such vehicles, so that they contribute to the costs of operating and maintaining the road system as well as contributing to revenues from general taxation, as do conventional vehicles.

♦ Public transport is an alternative to the car for many journeys. Its attraction is increased if rail and buses services are frequent, well integrated, with cashless ticketing and online journey planning. A high-quality public transport service is important for the quality of life in cities, and in towns and rural areas for those without access to a car. Public subsidy is needed to sustain such services.

♦ The main impetus for changed travel behaviour is the need to decarbonise the transport system. At present, the relative contributions needed from new technologies and behavioural change are unclear. The UK government's commitment to decarbonise motor transport is ambitious. But changes in charges and subsidies may also be needed to keep on track to the Net Zero objective.

The use of fossil fuels for propulsion has been essential for the transport system we have inherited, but the implications for climate change are now well understood and fossil fuel use must end. We are now in the era of transport decarbonisation. Substantial changes in technology are needed, and probably also in our travel behaviour. Many remedies have been proposed, but there are no simple solutions – no rules of thumb as yet. Innovation is therefore urgent – innovation in technology, policy, planning, operations and governance – if we are to be able to retain the benefits of access to which we have become habituated.

Endnotes

Notes for chapter 1

1 Kantar, for the Department for Transport, Transport and Technology. 2019. Public attitudes tracker: Wave 4 summary report (September).
2 Chatterjee *et al.* (2019).
3 National Infrastructure Commission. 2021. The long term role of cars in towns. Report, July.
4 I discussed the health and technological aspects of pollutants from motor vehicles in some detail in chapter 1 of Metz (2019).
5 Holgate and Stokes-Lampard (2017).
6 The published papers of the Committee on the Medical Effects of Air Pollutants indicate the complexity of the task faced by environmental epidemiologists.
7 Greater London Authority. 2020. Central London Ultra Low Emission Zone – ten month report. Report, April.
8 OECD. 2020. Non-exhaust particulate emissions from road transport: an ignored environmental policy challenge. Report, 2020.
9 Seaton *et al.* (2005).
10 Highways England. 2021. Smart motorways: first year progress report. Report, March.
11 House of Commons Transport Committee. 2021. Rollout and safety of smart motorways.
12 Jessica Murray. 2020. How Helsinki and Oslo cut pedestrian deaths to zero. *The Guardian*, 16 March.
13 Melia (2015).
14 See http://drivingchange.org.uk/hen-and-stag-parties/.
15 Department for Transport. 2020. Future of transport regulatory review: summary of responses. Report, October.
16 Department for Transport. 2016. Evaluation of concessionary bus travel: the impacts of the free bus sass. Report, 2016.

Notes for chapter 2

1 For a fuller account of the evidence and implications, see Metz (2021a).
2 Metz (2021b).
3 Metz (2010).

4   Recall from school geometry that the area of a circle is proportional to the radius squared; here, the 'radius' is the distance travelled in the time available and the 'area' is a measure of the places that can be reached in this time.
5   Department for Transport. 2017. Journey time statistics 2017 (Table 0201).
6   Metz (2013).
7   Metz (2010).
8   Table 0701 from the 2017 National Travel Survey shows the distance travelled per person per year for a household with one car was 6,000 miles, whereas for a household with two or more cars it was 8,500 miles, indicating that a second or third car in a household is used less than the first car.
9   Transport for London. 2019/20. London Travel Demand Survey.
10   Chatterjee et al. (2018).
11   Stokes (2013).
12   Transport for London publishes an annual series of reports called 'Travel in London' that include extensive statistics.
13   Greater London Authority. 2019. Trend-based population projections.
14   Transport for London. 2019. Travel in London. Report 12, figure 9.2.
15   Metz (2015).
16   Jones (2018).
17   The Department for Transport's 'Appraisal and modelling route map' (July 2020) notes that the long-term assumption about population growth has been reduced from 0.3% per annum to 0.15% per annum over the next fifty years.
18   *The Economist*. 2020. Japanese abroad: the endangered tourist. Article, 29 February.

Notes for chapter 3

1   National Infrastructure Commission. 2021. Infrastructure, towns and regeneration. Report, September.
2   This example is taken from work commissioned by the Department for Transport. See Batley et al. (2019).
3   Department for Transport. 2021. Integrated Rail Plan for the North and Midlands. The quotation is from paragraph 2.8.
4   My critique of the orthodox approach to transport investment appraisal can be found in Metz (2017) and Metz (2021a).
5   Metz (2021c).
6   Metz (2006).
7   Highways England. 2020. Economic analysis of the second road period. Report, July.
8   Department for Transport. 2021. TAG unit A1.2, scheme costs (November).
9   Bain (2009).
10   Department for Business, Energy and Industrial Strategy. 2020. Energy White Paper: powering our Net Zero future. White Paper, December.
11   An example of the Department for Transport's prescriptive guidance is the eighty-three-page 'TAG unit M2.1 variable demand modelling'.

12 The most insightful book on transport modelling is Yaron Hollander's *Transport Modelling for a Complete Beginner* (2016), particularly the discussion of the limitations of modelling.
13 Department for Transport. 2019. NTM future development: quality report. Report, November.
14 Department for Transport. 2020. Full business case: High Speed 2, Phase 1. Report, April.
15 Department for Transport. 2021. Integrated Rail Plan for the North and Midlands. Report.
16 National Infrastructure Commission. 2018. National infrastructure assessment. Report, July. The NIC view has been endorsed by the March 2020 Centre for Cities report 'Getting moving – where can transport investment level up growth?'.
17 National Infrastructure Commission. 2020. Rail needs assessment for the Midlands and the North. Final report, December.
18 In 'Garden villages and garden towns – vision and reality' from 2020, the campaign group Transport for New Homes persuasively critiqued the practice of constructing new housing on greenfield sites with few alternatives to travelling by car.
19 Transport Commission South East Wales. 2020. Final recommendations. Report, November.
20 For a critique of the economic case, see New Economics Foundation. 2018. Flying low: the true cost of Heathrow's third runway. Report.
21 See http://drivingchange.org.uk/heathrow-at-the-end-of-the-runway/.
22 Department for Transport. 2014. Value for money assessment for cycling grants. Report, August.
23 Department for Transport. 2019. Reported road casualties in Great Britain: 2019 annual report.
24 See https://tfl.gov.uk/corporate/safety-and-security/road-safety/vision-zero-for-london.
25 Department for Transport. 2021. Vehicle speed compliance statistics for Great Britain: 2020. Report, 13 July.
26 RAC. 2020. Report on motoring 2020.
27 Department for Transport. 2021. Transport statistics Great Britain. Report, table TSGB1308 RPI.
28 Office for National Statistics. 2019. Household expenditure on motoring for households owning a car. Report.
29 Metz (2018).

Notes for chapter 4

1 I discussed these new technologies at greater length in my previous book, *Driving Change* (Metz 2019).
2 Castelvecchi (2021).
3 The Faraday Institution. 2020. The importance of coherent regulatory and policy strategies for the recycling of EV batteries. Report, September.

4   The International Energy Agency has published a comprehensive study titled 'The role of critical minerals in clean energy transitions' (May 2021).
5   BloombergNEF. 2021. Hitting the EV inflection point. Report, May.
6   This estimate is from the *Financial Times*'s Alphaville blog, April 2021.
7   Emily Nagler. 2021. Standing still. Report, July, RAC Foundation.
8   E. Birkett and W. Nicolle. 2021. Charging up. Report, Policy Exchange.
9   Transport for London. 2021. Travel in London. Report 14, p. 142.
10 International Council on Clean Transportation. 2020. Real-world usage of plug-in hybrid electric vehicles. Report, September.
11 Ofgem. 2021. Enabling the transition to electric vehicles: the regulator's priorities for a green fair future. Report, September.
12 Schaller (2021).
13 Erhardt *et al.* (2019).
14 Transport for London. 2019. Travel in London. Report 12.
15 Transport for London. 2016. Travel in London. Report 9, section 6.9.
16 Tirachini (2020).
17 Le Vine and White (2020, section 2.8.4).
18 Button (2020). See also commentary by Hubert Horan at www.nakedcapitalism.com/category/uber.
19 See www.gov.uk/government/publications/vat-and-the-sharing-economy-call-for-evidence.
20 High Court case nos CO/4087/2020 and CO3046/2021 (6 December 2021).
21 BritainThinks. 2021. Future of transport deliberative research. Report for Department for Transport, March.
22 Currie and Fournier (2020).
23 For a comprehensive consideration of MaaS issues, see International Transport Forum. 2021. The innovative mobility landscape: the case of Mobility as a Service. Policy Paper 92.
24 See, for example, Marsden *et al.* (2019).
25 Lagadic *et al.* (2019).
26 Moody *et al.* (2021).
27 Department for Transport. 2020. Road traffic statistics, Report, table TRA0102.
28 Department for Transport. 2020. Benchmarking minor road traffic flows for Great Britain: 2018 and 2019. Report.
29 Department for Transport. 2020. Vehicle licencing statistics. Report, table VEH0402.
30 Department for Transport. 2020. Road traffic estimates: Great Britain 2019. Report.
31 Derrow-Pinion *et al.* (2021).
32 Emmerson (2014).
33 Highways England. 2020. Economic analysis of the second road period. Report, July.
34 Thaler (2018).
35 For a fuller discussion of automated vehicles, see Metz (2019). See also the 2020 MIT research brief 'Autonomous vehicles, mobility and employment policy: the roads ahead' by J. Leonard, D. Mindell and E. Stayton. Tom Standage's *Brief History of Motion* (2021) has an accessible account of the development of AVs.

36   Patrick McGee. 2021. Has Big Tech backfired on robotaxis? *Financial Times*, 20 July.

Notes for chapter 5

1   Department for Transport. From 2020 (continuous series). Transport use during the coronavirus (COVID-19) pandemic. Statistics for car, cycle and public transport.
2   Court of Appeal citation number [2021] EWCA Civ 1197.
3   Civil Aviation Authority statistics.
4   For a discussion of possible future developments of Covid-19, see Telenti *et al.* (2021) and Murray (2022).
5   Transport for London. 2019. Travel in London. Report 12, section 14.5.
6   M. O'Connor and J. Portes. 2021. Estimating the UK population during the pandemic. Report, January, Economic Statistics Centre of Excellence.
7   Greater London Authority. 2021. Population change in London during the COVID-19 pandemic. Report, May.
8   Greater London Authority: see https://data.london.gov.uk/dataset/londons-population.
9   Tom Forth. 2021. Brownfield first! Blogpost, 17 October (www.tomforth.co.uk).
10   National Infrastructure Commission. 2021. Impact of historic shocks on infrastructure demand. Report, January.
11   Office for National Statistics. 2021. Homeworking in the UK labour market: 2020. Report, May.
12   Office for National Statistics. 2021. Business and individual attitudes towards the future of homeworking, UK. Report for the April to May period.
13   Frances Cairncross's book *The Death of Distance* was published in 2001.
14   Owen Walker and Joshua Franklin. 2021. UK bank office returns hang in the balance. *Financial Times*, 13 June.
15   Nicholas Bloom. 2021. Happier, more efficient: why working from home works. *The Guardian*, 22 March.
16   Gratton (2021).
17   Discussed by the Bank of England's chief economist Andy Haldane in a speech titled 'Is home working good for you?' on 14 October 2020.
18   Charted Management Institute. 2021. Making hybrid inclusive: key priorities for policymakers. Report, October.
19   Cited in Barrero *et al.* (2021).
20   Simon Samuels. 2021. Banks must heed the value of institutional memory. *Financial Times*, 1 July 2021.
21   Andrew Hill. 2021. The flexibility factor: who is going back to the office? *Financial Times*, 20 September.
22   George Hammond. 2021. The developers still betting on the London office market. *Financial Times*, 17 May.
23   KPMG. 2021. New working patterns and the transformation of UK business landscape. Report, September.
24   Georgia Lowe. 2021. Business travel during Covid-19: a survey of UK businesses. Report, Ipsos MORI for the Department for Transport.
25   The Centre for Cities. 2021. High streets recovery tracker. Report.

26 National Travel Survey, table NTS0403 (2021).
27 Office for National Statistics. 2020. Impact of the coronavirus (COVID-19) pandemic on retail sales in 2020. Report.

Notes for chapter 6

1   Transport for London. 2013. London's street family: theory and case studies. Report.
2   International Transport Forum. 2021. Reversing car dependency: summary and conclusions. Round table report 181.
3   Kantar for the Department for Transport. 2020. Public opinion survey of traffic and road use: general public research. Report, November.
4   Kantar for the Department for Transport. 2021. Low Traffic Neighbourhoods residents' survey. Report, January.
5   See www.ealing.gov.uk/ltnresults.
6   OnLondon. 2021. Ealing: what next for Low Traffic Neighbourhoods? Article, 14 October (www.onlondon.co.uk).
7   Transport for London. 2021. Travel in London. Report 14, p. 124.
8   Department for Transport. 2021. Gear change: one year on. Report.
9   The most comprehensive collection of evidence is that in Cairns et al. (2002), with further evidence in the reference at endnote 2 above.
10 Greater London Authority. 2020. Central London Ultra Low Emission Zone – ten month report. Report, April.
11 C40 Cities Knowledge Hub. 2021. Introducing spotlight on 15-minute cities. Report, May.
12 For sceptical views on 15-minute cities, see www.lse.ac.uk/Cities/urban-age/debates/key-takeaways-3.
13 Transport for London. 2021. Travel in London. Report 14, p. 225.
14 Juliana O'Rourke. 2021. Why 15 minute cities are a good idea. *Local Transport Today*, issue 823, 17 May.
15 Transport Planning Society. 2020. Annual members survey 2019/20.
16 City of Copenhagen. 2019. The Bicycle Account 2018: Copenhagen City of Cyclists. See also Transport for London, Travel in London Report 12.
17 Kodukulo et al. 2018. Living moving breathing: ranking of European cities in sustainable transport. Report, Wuppertal Institute.
18 Buehler et al. (2017).
19 International Energy Agency. 2020. Share of SUVs in total car sales in key markets, 2010–2019. Report.
20 New Weather Institute. 2021. Mindgames on wheels: how advertising sold false promises of safety and superiority with SUVs. Report, April.
21 MIT Energy Initiative. 2019. Insights into Future Mobility. Report, section 3.4.
22 Mattioli et al. (2016).
23 Lynn Sloman et al. 2018. Impact of the local sustainable transport fund: synthesis of evidence. Report to Department for Transport.
24 Lynn Sloman et al. 2019. Summary and synthesis of evidence: cycle city ambition programme 2013–2018. Report to Department for Transport.
25 Michie et al. (2011).

26 Petersen (2016).
27 Transport for London. 2021. Financial sustainability plan. Report, January.
28 Buehler *et al*. (2017).
29 National Infrastructure Commission. 2020. Rail needs assessment for the Midlands and the North: final report. Report, December.
30 S. Jeffrey and K. Enenkel. 2020. Getting moving: where can transport investment level up growth? Report, March, Centre for Cities.
31 What Works Centre for Local Economic Growth. 2015. Evidence review 7: transport. Report.
32 I. Docherty and D. Waite. 2018. Evidence review: infrastructure. Report, Productivity Insights Network.
33 Marshall and Dumbaugh (2020).
34 HM Government. 2022. Levelling up the United Kingdom (CP 604). White Paper, February.

Notes for chapter 7

1   Climate Change Committee. 2020. Sixth carbon budget. Report, December.
2   Department for Transport. 2021. Decarbonising transport. Policy Paper, July.
3   HM Government. 2021. Net Zero strategy: build back greener. Policy Paper, October.
4   Department for Transport. 2009. Low carbon transport: a greener future (July, figure 2.7).
5   National Travel Survey, table 0702 (2020).
6   Enrica Papa. 2020. Would you ditch your car if public transport were free? *The Conversation*, 5 March.
7   Centre for London. 2020. Reclaim the kerb: the future of parking and kerbside management. Report, March.
8   Department for Transport. 2021. Decarbonising transport (July, p. 7).
9   Department for Transport. 2018. Road traffic forecasts 2018. Report, scenario 7.
10 Policy Exchange. 2021. Great restorations: setting the long-term direction for climate and environmental policy. Report.
11 European Commission. 2021. Questions and answers – emission trading – putting a price on carbon. Report, 14 July.
12 The Department for Transport commissioned a technology report from Mott Macdonald *et al*. which was titled 'Decarbonising UK transport (final report and technology roadmaps)' (March 2021).
13 Competition & Markets Authority. 2021. Electric vehicle charging market study. Final report, July.
14 Government Response to the Climate Change Committee's 'Progress in reducing emissions – 2021 report to Parliament'.
15 See note 12.
16 House of Commons Library. 2021. Aviation, decarbonisation and climate change. Report, 20 September.
17 UK Parliament POST Note 616, 2020.
18 International Energy Agency. 2020. Tracking report: aviation (June).

19   Climate Change Committee. 2020. Sixth carbon budget. Methodology Report, December.

20   International Council on Clean Transportation. 2020. Vision 2050: a strategy to decarbonize the global transport sector by mid-century. Report.

21   Brand *et al.* (2020).

22   Climate Assembly UK. 2020. The path to Net Zero. Report, September.

23   House of Commons Transport Committee. 2021. Inquiry into zero emission vehicles and road pricing. Written evidence submitted by Professor Jillian Anable (EVP0113).

24   Greater Cambridge Citizens' Assembly on Congestion, Air Quality and Public Transport. Final report, November 2019.

25   Greater Cambridge Partnership. 2021. Making Connections public consultation, Autumn.

26   Adair Turner. 2020. Techno-optimism, behaviour change and planetary boundaries. Keele World Affairs Lectures on Sustainability, 12 November.

References

Bain, R. (2009). Error and optimism bias in toll road traffic forecasts. *Transportation* **36**, 469–482.

Barrero, J., Bloom, N., and Davis, S. (2021). Why working from home will stick. Working Paper 28731, National Bureau of Economic Research.

Batley, R., *et al.* (2019). New appraisal values of travel time saving and reliability in Great Britain. *Transportation* **46**, 583–621.

Brand, C., Anable, J., Ketsopoulou, I., and Watson, J. (2020). Road to zero or road to nowhere? Disrupting transport and energy in a zero carbon world. *Energy Policy* **139**, article 111334.

Buehler, R., Pucher, J., Gerike, R., and Götschi, T. (2017) Reducing car dependence in the heart of Europe: lessons from Germany, Austria, and Switzerland. *Transport Reviews* **37**(1), 4–28.

Button, K. (2020). The 'Ubernomics' of ridesourcing: the myths and the reality. *Transport Reviews* **40**(1), 76–94.

Cairns, S., Atkins, S., and Goodwin, P. (2002). Disappearing traffic? The story so far. *Proceedings of the Institution of Civil Engineers – Municipal Engineer* **151**(1), 13–22.

Castelvecchi, D. (2021). Electric cars and batteries: how will the world produce enough? *Nature* **596**, 336–339.

Chatterjee, K., *et al.* (2018). Young people's travel – what's changed and why? Review and analysis. Report to the Department for Transport, University of the West of England.

Chatterjee, K., *et al.* (2019). Access to transport and life opportunities. Report to the Department for Transport, NatCen Social Research.

Currie, G., and Fournier, N. (2020). Why most DRT/micro-transits fail. What the survivors tell us about progress. *Research in Transportation Economics* **83**, article 100895.

Derrow-Pinion, A., *et al.* (2021). ETA prediction with graph neural networks in Google Maps. Preprint, available at https://arxiv.org/abs/2108.11482.

Emmerson, G. (2014). Maximising use of the road network in London. In *Moving Cities: The Future of Urban Travel*, ed. S. Glaister and E. Box. London: RAC Foundation.

Erhardt, G., *et al.* (2019). Do transportation network companies decrease or increase congestion? *Science Advances* **5**(5), article eaau2670.21.

Gratton, L. (2021). How to do hybrid right. *Harvard Business Review*, May–June, 66–74.

Hensher, D. A., Ho., C, Reck, D., Smith, G., Lorimer, S., and Lu, I. (2021). The Sydney Mobility as a Service (MaaS) trial: design, implementation, lessons and the future. Report, University of Sydney.

Holgate, M., and Stokes-Lampard, H. (2017). Air pollution – a wicked problem. *British Medical Journal* **357**, article j2814.

Jones, P. (2018). Urban mobility: preparing for the future, learning from the past. CREATE Project Summary and Recommendations (www.create-mobility.eu).

Lagadic, M., Verloes, A., and Louvet, N. (2019). Can car sharing services be profitable? A critical review of established and developing business models. *Transport Policy* **77**, 68–78.

Le Vine, S., and White, P. (2020). The shape of changing bus demand in England. Report, Independent Transport Commission, London.

Marsden, G., Anable, J., Bray, J., Seagriff, E., and Spurling, N. (2019). Shared mobility: where now? Where next? The second report of the Commission on Travel Demand. Report, Centre for Research into Energy Demand Solutions, Oxford.

Marshall, W., and Dumbaugh, E. (2020). Revisiting the relationship between traffic congestion and the economy: a longitudinal examination of US metropolitan areas. *Transportation* **47**, 275–314.

Mattioli, G., Anable, J., and Vrotsou, K. (2016). Car dependent practices: findings from a sequence pattern mining study of UK time use data. *Transportation Research Part A* **89**, 56–72.

Melia, S. (2015). *Urban Transport Without the Hot Air*. Cambridge: UIT.

Metz, D. (2006). Accidents overvalued in road scheme appraisal. *Proceedings of the Institution of Civil Engineers: Transport* **159**(4), 159–163.

Metz, D. (2010). Saturation of demand for daily travel. *Transport Reviews* **30**(5), 659–674.

Metz, D. (2013). Mobility, access and choice: a new source of evidence. *Journal of Transport and Land Use* **6**(2), 1–4.

Metz, D. (2015). Peak Car in the Big City: reducing london's greenhouse gas emissions. *Case Studies on Transport Policy* **3**(4), 367–371.

Metz, D. (2017). Valuing transport investments based on travel time saving: inconsistency with United Kingdom policy objectives. *Case Studies on Transport Policy* **5**(4), 716–721.

Metz, D. (2018). Tackling urban traffic congestion: the experience of London, Stockholm and Singapore. *Case Studies on Transport Policy* **6**(4), 494–498.

Metz, D. (2019). *Driving Change: Travel in the Twenty-First Century*. Newcastle: Agenda Publishing.

Metz, D. (2021a). Time constraints and travel behaviour. *Transportation Planning and Technology* **44**(1), 16–29.

Metz, D. (2021b). Plateau car. In *International Encyclopedia of Transportation*, ed. Roger Vickerman, volume 6, pp. 324–330. Elsevier.

Metz, D. (2021c). Economic benefits of road widening: discrepancy between outturn and forecast. *Transportation Research Part A* **147**, 312–319.

Michie, S., van Stralen, M., and West, R. (2011). The behaviour change wheel: a new method for characterising and designing behaviour change interventions. *Implementation Science* **6**, 42.

Moody, J., *et al.* (2021). The value of car ownership and use in the United States. *Nature Sustainability* **4**, 769–774.

Murray, C. (2022). COVID-19 will continue but the end of the pandemic is near. *Lancet* **299**, 417–419.

Petersen, T. (2016). Watching the Swiss: a network approach to rural and exurban public transport. *Transport Policy* **52**, 175–185.

Schaller, B. (2021). Can sharing a ride make for less traffic? Evidence from Uber and Lyft and implications for cities. *Transport Policy* **102**, 1–10.

Seaton, A., *et al.* (2005). The London Underground: dust and hazards to health. *Occupational and Environmental Medicine* **62**, 355–362.

Stokes, G. (2013). The prospects for future levels of car access and use. *Transport Reviews* **33**(3), 360–375.

Telenti, A., *et al.* (2021). After the pandemic: perspectives on the future trajectory of COVID-19. *Nature* **596**, 495–504.

Thaler, R. (2018). Nudge, not sludge. *Science* **361**, 431.

Tirachini, A. (2020). Ride-hailing, travel behaviour and sustainable mobility: an international review. *Transportation* **47**, 2011–2047.

Index